ONE KISS OR TWO?

ONE KISS OR TWO?

THE ART AND SCIENCE OF SAYING HELLO

ANDY SCOTT

Overlook Duckworth
New York • London

This edition first published in the United States and the United Kingdom in 2017
by Overlook Duckworth, Peter Mayer Publishers, Inc.

NEW YORK
141 Wooster Street
New York, NY 10012
www.overlookpress.com
For bulk and special sales please contact sales@overlookny.com,
or to write us at the above address.

LONDON
30 Calvin Street, London E1 6NW
T: 020 7490 7300
E: info@duckworth-publishers.co.uk
www.ducknet.co.uk
For bulk and special sales please contact sales@duckworth-publishers.co.uk,
or write to us at the above address.

Cataloging-in-Publication Data available from the Library of Congress

A catalogue record for this book is available from the British Library

Text design and typesetting by Tetragon, London

978-1-4683-1601-8 (US)
978-0-7156-5183-4 (UK)

1 3 5 7 9 10 8 6 4 2

Contents

For my friends and family – finally

Preface

I was standing outside a hotel in Istanbul, trying to get my bearings, when the idea for this book first came to life. Two bouncer-like figures stood either side of the entrance looking tough. It was hardly the most welcoming sight and I held back from asking for directions. But, then, to my amazement, a young guy wandered over and greeted them with double kisses, lightly holding their chins as he passed from cheek to cheek. Suddenly, the stone-faced doormen were all smiles and laughter. It was an intimate scene and I was left wondering why, back home in the UK, we sometimes struggle with even a handshake. What was so different in Turkey?

In different forms, this was a question that stayed with me over the following months, as I made my way across Europe and down Africa. I'd finally finished my PhD and was on a giant trip, trying to get from Cambridge to Cape Town overland using public transport. As I crossed from country to country and culture to culture, I began to keep note of the different greetings I encountered. I was particularly struck by the reception I received in Sudan. It was a country I was nervous about entering, with its news dominated by war and instability. But, where I was travelling, I needn't have worried. Apart from the barren landscape and burning heat, my overwhelming impression was of incredible friendliness. If in doubt, all I had to do was use the traditional Arabic greeting *As-salamu alaykum* ('Peace be upon you'), and people's faces would light up as they returned the gesture. Back home, it was a sentiment reserved for my yearly visit to church and even then was a bit embarrassing, but here it felt like the perfect way to break down any barriers or suspicion, and I found myself greeting like never before. The physical side of Sudanese greetings also had a special quality. Between men, they often began with an exaggerated pat to the right shoulder before combining this with a handshake. I started off tentatively, but there was always particular delight when I joined in the ritual.

In Ethiopia, the shoulder pat became the shoulder bump, which was then replaced by a sort of finger snap in parts of Kenya. Of all the greetings I encountered, though, my favourite was towards the end of my trip, in Namibia. Having just arrived, I was standing in a bank changing some money, when I noticed two men approaching each other, smiling broadly. The younger man stopped a couple of metres short and crouched down, holding his hands together as if about to pray. He then gave a short burst of applause, while the other man watched in appreciation. No one else seemed to notice, but I was transfixed. The whole display finished with a hug and a handshake, accompanied by a warm and enthusiastic exchange. Again, what was different in Namibia, I wondered? How did people come to greet each other like superstars?

Back home, inspired by my trip, the idea of a book on greetings took hold. 'But hasn't it been done before?' people asked. It seemed so obvious. After all, there are books on just about every aspect of human behaviour. I'd even seen books on queuing. *Surely there must be one on the first moment of interaction.* I nervously turned to Google, certain I would find the book that would kill the idea. But, to my relief, all I found, buried among a bunch of sites selling Christmas cards, was a children's picture book which, while fun, didn't give greetings the treatment I'd begun to think they deserved.

At first, I thought in terms of a global guide, and began to search for and catalogue different greetings from around the world. There were more variations of the handshake, such as the diagonal clasp and fist bump, and then different twists on our most common greetings. But I was determined to find the most unusual and reveal humans in all their oddness. It was like being on an Indiana Jones adventure, albeit sitting in front of a computer screen. There are the more well-known ones, such as the Maori nose rub (the *hongi*) in New Zealand or the prayer sign (the *namaste*) in parts of Asia. But, as I got deeper, past the first pages, I started to discover more outlandish customs – behaviour that made me question what it really means to say hello. There's cheek-sniffing in Tuvalu and foot-kissing in Bangladesh; in Tibet, there's the practice of poking tongues out, while members of the Maasai tribe in Kenya mark each other's arrival by spitting on the floor. One article even described the convention among Central Eskimos of slapping strangers around the face when first meeting. The stranger is then entitled to slap back and more rounds follow, in ever-greater intensity, until someone finally falls to the floor, presumably in a lot of pain.

All of this challenged my understanding of what it means to greet. Being a naturalist at heart, I started to look beyond our own species, wondering if and how the rest of the animal world says hello. My own knowledge mostly came from watching our cats. I realise that there's a danger of thinking of our pets in human terms, but I was struck by how they seemed to have their own special greetings. Of the current bunch, Alice (who is sitting next to me as I type) has a peculiar way of quivering her tail whenever she walks in, which usually lasts for a few seconds until she's had her first stroke. Then there's her brother, Bedford, who is much more vocal, especially when bringing in mice or birds, which he announces with a deep meow, as if he's summoning the spirits. Towards each other, they mostly hiss and growl. But at friendlier moments they'll rub noses and inspect each other's behinds. I'd even heard of cats headbutting each other. Did these count as greetings, I wondered? And what about species closer to our own?

In the end, as I collected all of the different customs, I realised that my global-guide idea had turned into something else. I'd started to see these first moments of interaction as a window into different cultures, maybe even our species as a whole. The thing that began to interest me was not so much what people did but understanding why they did them. How did these greetings come about, evolve and spread? How did we come to shake hands, pat shoulders and slap faces? And why do we even bother with these rituals in the first place? Were they simply a matter of culture or, looking at our cats, was our biology involved? Stepping back, I was struck by a simple fact: that greetings are at once both universal and particular; that is, they're something we all do but each in our own special way. While basic, I felt like this touched on something profound – something about what it means to be human and about our place in the world.

So these are the aims of this book: to reveal greetings in all their variety and complexity, and to understand their origin, why we do them and how they have evolved. To this end, what follows takes us on a journey in search of answers to some basic questions. Starting out, though, we'll look at where and how our greetings can go wrong. In truth, the driving inspiration behind this book was not so much travelling the world and marvelling at humanity's diversity as being at home and getting in a constant muddle. As we'll see, our greetings have become a big source of confusion and embarrassment, and things seem to be getting worse. Part of the problem is that these first moments of interaction are important, though we don't always know why or how. In subsequent chapters, we'll try to get to grips with some of this, to see if we can help get our greetings right or at least better understand why we get them wrong. First, we'll dissect what's in a greeting to find where the problems lie. We'll then see if there really is such a thing as a perfect greeting and how we can know. Next, we'll tackle the issue of handling greetings from around the world to understand how we should behave when meeting people from different cultures. Having come to terms with our differences, though, we'll go in search of what unites us and whether our greetings share any common origins. Here, we'll step back into our evolutionary past and explore our connections with the rest of the animal world. In doing so, we'll see how our strangest greetings actually reveal some fundamental aspects of human behaviour that connect us all. We'll also look more closely at some of the physiological dimensions to our greetings, exploring how our senses and, most importantly, our brains come into play. Having dug into the origins of our greetings, we'll trace how they've evolved and spread, using them as a barometer for our entire history. Finally, we'll take a look into our future, assessing where our greetings are going.

All in all, stretching across the whole of human history, searching every corner of the planet and bringing in the rest of the animal world, it's going to be an interesting journey. Along the way, we'll meet and be guided by some of the foremost experts on human behaviour. We'll get advice from etiquette coaches and body-language gurus who have made a career out of telling the rest of us how to behave. But to get at some of the bigger questions we'll need to bring in the academics. We'll meet a pioneer of the study of human interaction, as well as leading sociologists and anthropologists who can explain what connects us, as well as our differences. In searching for the origins of our greetings, we'll be helped by zoologists and palaeontologists who have spent their lives studying different animals and our earliest ancestors – people who have changed our view of what it means to be human. We'll also meet pioneering neuroscientists who'll show us what's going on inside our brains, revealing aspects of our behaviour that we're not even aware of. And to get a sense of what's changed and where we're heading, we'll see what the historians have to say and meet someone who claims to predict the future.

As far as my own background goes, while I studied history, I don't have any formal training in any of these areas. But, in some ways, this works to my advantage. Coming to the subject of greetings without any specialisation or academic loyalties, I wanted to use it as a chance to look across all of the fields, as much as possible pulling their insights together. Starting out, the only conviction I did have was that, when it comes to explaining human behaviour, there can be no other way.

As well as drawing on the experts, though, I've also done some of my own primary research, even if only to test their conclusions. Here, I've been inspired by the notions of 'small science' and 'armchair anthropology'. Spurning the need for big laboratories, vast teams and pots of money, 'small science' embraces the study of the everyday. It's science at its most demo-cratic – armed with enough curiosity and a critical eye, anyone can have a go. After all, when it comes to human behaviour, we're all scientists trying to work out its causes.

As for 'armchair anthropology', it goes back to the days of the early Victorian anthropologists and their tendency to rely on the second-hand accounts of explorers, traders and colonialists to explain different cultures. Today, though, I think it can be used in a more positive way. With the invention of television and the internet, it really is possible to explore the world and different cultures in a meaningful way without leaving the comfort of your armchair. I might not be able to meet tribes for the first time, but I can watch people who have. And the explosion of reality TV, however tedious at times, gives us the chance to watch people going about their daily lives.

But I've also got out of my chair and followed the example of modern anthropologists, using their key method: 'participant observation'. Essentially, this is understanding different cultures by experiencing them. Typically, the intrepid researcher would head off to stay with some remote tribe, spending months, even years, recording and experiencing all aspects of their lives. Having originally had grand ideas about travelling the world to try all the different greetings, I've had to limit my horizons. In this respect, though, I'm lucky to live in London, since I have the world on my doorstep. Today, with its 8.7 million inhabitants speaking more than 300 languages, it's one of the most multicultural cities in the world, allowing me to explore and experience customs from all corners of the globe, as well as my own.

What follows, then, is the story of my search for answers to some basic questions. Inspired by my own travels and struggles, it describes my journey across different disciplines, meeting the experts and collecting insights from around the world and everyday life. All this will shed light on our greetings, helping us to understand what makes them important and how to get them right, but it will also open up some bigger questions – questions about what it means to be human and our place in the world. The Scottish naturalist John Muir best captures the spirit in which I have approached the book: 'When we try to pick out anything by itself, we find it hitched to everything else in the universe.' When it comes to greetings, I hope that, by the end, you will know what I mean.

1

It's a Minefield Out There

I was propped up against a bar in London, waiting for my blind date, Penny, to show up. My nerves were building – all being well, this would be the person I'd spend the rest of my life with. But I was also feeling uncertain about the first moment of interaction, whether to go for a handshake, hug or kiss. In the end, I decided on a single kiss. A handshake seemed too formal, as if I was turning up for a job interview (even though it felt like I was); a hug might suggest that I'd already settled for 'just friends'; while a double kiss, although the norm, still seemed pretentious in my eyes. Finally, a girl walked in fitting Penny's description and smiled. I stepped forward and found myself offering my hand while simultaneously going to give her a kiss on the cheek. Happily, she reciprocated. But as I pulled back, Penny was left hanging, waiting for a second. I found myself leaning back in, but it was too late and I caught the corner of her mouth as she pulled away. We laughed it off, but my nerves went into overdrive and I found it hard to relax. There was no second date.

There's no doubt: greetings can be awkward – fraught with doubt and embarrassment about when, who and how to greet. And, sadly, like a lot of things in life, it's the times that have gone wrong that can leave their mark. From my earliest memories, greetings just seemed unnecessary and vaguely unpleasant, something adults did and I was pushed into. I'd wriggle free from any hug and wipe away the kisses so I could get on with the serious business of playing. But slowly, imperceptibly, things began to change as I became a teenager, and there came a time when it was no longer enough to be a bystander in these exchanges. With a little help from my brothers, I developed a passable handshake, which began to feel normal. Girls, though, were another matter. For whatever reason, a handshake didn't feel quite right and certainly wasn't cool. Slowly, I began to manage a hug with my friends and cousins, but it was excruciating stuff. Just as I was beginning to feel a little more self-assured, I headed off to university, and suddenly encountered a whole new level of confidence, as I mixed with people from private schools for the first time. Their handshakes were firmer and 'hellos' more assertive. But, most unsettling, a few were using the double kiss, even adding a 'mwah, mwah'. They seemed a bit ridiculous, if not completely pretentious, and the inverted snob in me resolved not to do them. But resistance was futile, as the double kiss spread like a virus. There came the day when, against all instinct, I became a double-kisser.

Now, with the full arsenal of greetings at my disposal, I was baffled. It was a minefield, knowing which one to go for. But, from what I'd seen at university, it seemed that the key was to be confident or, if in doubt, to follow the person in front. Unfortunately, though, just when I thought I'd cracked it, my new strategy took a fatal knock. It was the summer after my first year at university and I was in London with my brother, Chris, to meet our older sister, Lizzie, and her friend, Sarah. Walking out of the Tube down to the Thames, I even joked about how greetings had become so awkward, with Sarah presenting a perfect example. For all our laughing, I was actually feeling uneasy about it. But I just had to be confident and, as the younger brother, could follow Chris's lead anyway. As we approached the river and spotted my sister (I can still remember the exact spot), I slid behind Chris, hoping he'd go for a handshake. But he'd met Sarah before and went straight in for a hug and kiss.

'And this is my brother, Andy,' Lizzie said.

I felt sick. *Don't be so pathetic*, I told myself, as I took charge of the situation and followed Chris's lead. But as I went in for the kiss, Sarah pulled back.

'Blimey,' she said, as I caught her cheek, 'he's confident'.

And that was it. Any true confidence in me was shot to pieces. I felt like some cocky upstart, like I'd violated some hidden rule of social behaviour. For the rest of the day, I held back, unable to make eye contact. My whole greetings strategy was in tatters.

I'm not sure I've ever fully recovered from that moment, but as I've got older and a little less self-conscious, most greetings have an established routine, which helps to ease the uncertainty. Still, though, there's often an element of doubt when patterns get interrupted. And now there are even more categories of people to greet, whether nieces and nephews, friends' children or pro-spective employers and colleagues. As ever, the most uncertainty comes with members of the opposite sex – knowing whether to hug and how many kisses to give, and even then whether you should actually make contact. First dates are especially difficult. Meeting your potential partner in life is nerve-wracking enough without there being extra doubt over the first interaction. But often worst are the more innocuous encounters, with people who straddle the line between being a mere acquaintance and a friend, or with someone you're meeting for a second time. A handshake can feel too formal while a hug might seem over-familiar. And then indecision turns to embarrassment when you realise that you've gone for something different to your fellow greeter. We do our best to shake these moments off but the awkwardness can infect the subsequent interaction. As the Irving Berlin lyric goes, 'The song is ended, but the melody lingers on.'

All of this is bad enough at home, but going abroad it gets worse. It's one thing visiting a country and getting a kick out of the various cultural quirks, but stay anywhere long enough and sooner or later you want to fit in. My first real experience of this was on my French exchange at the age of fourteen, when I was puzzled by the constant handshaking and had my first encounter with the double kiss. Years later, representing my school on a visit to Japan, I was expected to put on a respectable face, but found it difficult to bow without smirking. Going to live overseas for the first time, in Canada, even the once-reliable handshake started to cause problems. Suddenly, guys were going in diagonal or straight for the hug. And it took me weeks to realise that the constant refrain 'What's up?' had nothing to do with my demeanour and that the only response was 'Nothing much'.

More recently, I returned to Sudan, this time as a diplomat. Greetings were now part of my job, though there was no special training. Among the Sudanese, I mastered the shoulder pat, and got used to the constant enquiries about my family's health. The real issue, though, was among the rest of the international community. I'd fully expected a bit of double-kissing – after all, being a diplomat can be like living in a giant role play – but, amazingly, the *triple* kiss had become the norm. Whatever nationality, everyone was at it. Again, I tried to resist. Two was bad enough; the third just seemed ironic. But, again, resistance was futile. As I discovered, though, it didn't go down so well among my Sudanese colleagues.

For all my travels abroad, far from becoming a more confident and cosmopolitan greeter, I've become more confused than ever. At times, I've wondered if it's just me. But as I shared my experiences with others and started to look into the issue, I found that I'm not alone. In fact, everyone, it seems, has a story to tell. From tortured tales of half-kisses with the boss to intercultural fumbling, everyone, it seems remembers a bad greeting. And many people are similarly mystified by these basic social interactions, unsure when and how to say hello. Without doubt, social kissing is a major culprit. As the journalist Tom Geoghegan has put it, one kiss or two is 'the unspoken dilemma dividing Britain'.[1] And even the handshake is causing problems: a recent survey in Britain found that over two-thirds of people (76 per cent in my home region, East Anglia) are experiencing a crisis of confidence about whether they are doing it properly.[2] The problem was worse than I'd thought.

Could it be that all of this was somehow uniquely British? After all, we're renowned for being aloof and awkward. Attempting to shed light on our condition, a recent TV series based on a popular Twitter account set out to identify what it called 'VBPs' (Very British Problems).[3] Different pundits grappled with our various social shortcomings, from our inability to make small talk (unless it's about the weather) and deep fear of strangers to our horror at the thought of making a scene and special ability to avoid people. As one commentator put it, in Britain, 'the mark of a lunatic is getting to know everyone on your street.' The topic of greetings provoked a particular dread. 'The Brits can't greet each other,' was the basic conclusion. There's no doubt, then, that part of my own difficulties stems from where I'm from.

But it turns out that even the French, the biggest hand-shakers and kissers of all – and who might be blamed for much of our predicament – have difficulties. For some, the social kiss (*la bise*) has become a source of anguish rather than pride. And, while the British might agonise over whether it's one kiss or two, in France it's anywhere up to five. There's even a kissing map. In the United States, where everyone can seem so confident in their social interactions, there's been a recent backlash against all the hugging in high schools, while the new trend in some workplaces to kiss on the lips is making many people shudder. In Germany, one etiquette society has become so bothered by all the kissing at work that they've called for a complete ban, regarding it as a 'form of terror'.[4] And it's not just a problem of an uptight West. My Sudanese friends tell me that knowing when and with whom to pat shoulders can be a major cause of confusion, while an old housemate from India told me that she struggled growing up, particularly when it came to kissing her grandmother's foot. Perhaps, then, the main thing that distinguishes us Brits is that we've learned to celebrate being awkward.

Moreover, it's not just a question of being especially shy or introverted. Our entertainers, politicians and business leaders – those who might be regarded as our professional greeters – are suffering too. Watch the start of any international summit and it can seem like some strange insect mating ritual, as leaders advance, withdraw and sidestep, unsure of their next move. For them, it's even worse, as their gaffes are caught on camera and immediately broadcast across the world, preserved on YouTube for the rest of us to laugh at. Having spent a lot of time studying Anglo-American relations, one of my personal favourites was when Gordon Brown and George W. Bush met in Northern Ireland in June 2008 for their first summit. A grinning Brown stepped forward and went for the regular handshake, but at the last second Bush, also smiling, went in diagonally for a hip-hop-style clasp. The result was an awkward tangle, with three of

Brown's finger's sliding up Bush's shirt. Back home, the prime minister, who was already struggling, was widely mocked for 'losing his grip'. If two world leaders can't get it right, what hope is there for the rest of us?

Through the forces of globalisation, culture clash is for many of us an everyday occurrence. Technology is also playing a part, changing the way we communicate faster and demanding new forms of greeting. And, in recent years, people-watchers in the UK and the United States have observed a 'greetings inflation', an arms race of lips and limbs, whereby our ways of saying hello are getting more intimate: where a nod was once enough, we now shake hands; where we shook hands, we now hug; where we hugged, we now add a kiss, and so on. No one is quite sure why, but it's creating extra expectation, compounding our confusion and unease.

Mostly, we laugh off moments of awkwardness and take pleasure in seeing our leaders mess up. But fear of getting it wrong can be deeply unnerving, even paralysing. 'I feel permanently uncomfortable these days,' complains a high-flying City lawyer in an internet post, reflecting the mood of our times. Apprehension about saying hello has become so widespread that it could be regarded as a social condition. One of my friends even suggested a name: 'greetings anxiety'.[5] As journalist Shane Snow says, describing how he often gets in a fix about which greeting to go for: 'The more I think about it, the more I spiral counter-clockwise down the toilet of anxiety.'[6]

All of this was recently brought into sharp relief at a charity sale organised by one of my nieces. It was the perfect storm. There were a whole range of ages, a mix of friends and friends of friends, as well as all categories of relatives and distant relatives. It really was a minefield out there. But as I circled the edges,

I realised that it wasn't just me who was avoiding eye contact. The place was littered with half-kisses and aborted hugs. I stood behind the cake stand with my other niece and nephew, out of harm's way. They were finding it all similarly difficult. 'It's a nightmare,' they said in unison. It was then that it hit me: rather than opening up our interactions, greetings have become a *barrier* – the very act of saying hello has become a reason not to say hello. Something needed to be done.

All of this got me thinking: could I do something to help or at least explain? Could I somehow remove the doubt, perhaps uncovering some underlying rules to guide us through this social minefield, both at home and abroad? In trying to overcome my own difficulties, could I offer some small comfort to humanity? And how can something so familiar be so awkward? Where does all of the confusion come from? Do other animals share our uncertainty? Perhaps awkwardness, far from being a uniquely British trait, is a defining characteristic of our species. As co-founder of the Awkwardness Appreciation Society, I found it a satisfying thought.

The big issue in all of this is that greetings matter. Whether a casual nod, twenty-one-gun salute or sticking 'Hi' at the start of an email, our exchanges all start with some recognisable sign. In the extreme, a handshake between leaders can symbolise the opening of relations between two countries. For the rest of us, they often help to break the ice and put us at ease. As the wildlife cameraman Gordon Buchanan commented as he approached a remote village in the Amazon, feeling nervous: 'All you need is a warm welcome and a smile.'

Greetings reveal something about our relationships, conveying our feelings and status. Some of this comes through the particular greeting we choose, but also significant are the subtle variations, such as the length and strength of a hug. Perhaps greetings can even tell us something about people individually: not just where they're from, but their character. I'd read newspaper articles suggesting that you can tell if someone is conscientious and competent, even the state of their health, just by their handshake. A business leader even claimed that he could decide whether he was going to hire someone by the strength of their grip alone. On the other hand, concerned about hygiene, Donald Trump has described the handshake as 'barbaric'.

We're constantly told how much first impressions count, that we judge people within seconds of meeting them. President Kennedy commissioned an entire study into how he should greet other leaders to make the best impression. But how much do these first impressions really count? And how do our greetings play into them? Can we really judge people by the way they greet, even by the strength of their handshake? Take the story of Mark Latham, the challenger in Australia's 2004 presidential race. Going into the election, he was way ahead in the polls. But all this changed following his encounter with incumbent John Howard at a radio station on the eve of polling day. Meeting Howard, Latham pulled him in close and arched over his opponent, shaking his hand vigorously. Footage of the handshake quickly spread over the internet. The following day, Howard was re-elected, with polls suggesting that the main reason for the sudden swing had been the negative reaction to Latham's domineering handshake. Something had deeply offended the Australian public. And if a relatively ordinary greeting could have such an adverse effect, maybe they could work the other way too – maybe they really could win you votes or get you a job.

That our greetings matter might help explain some of our anxieties; it might also shed light on why we do them in the first place and what's really going on when we say hello. And, in turn, understanding why we greet might just help us know how to greet each other. To paraphrase Nietzsche, when we understand the why, the how becomes easy. Armed with the right information, maybe we can learn to make the perfect impression, ease our worries and avoid losing elections.

In his book *Queuing for Beginners*, the cultural historian Joe Moran notes how he was inspired by the French writer Georges Perec. Living in Paris in the 1960s, Perec encouraged his readers to look again at the things we find ordinary and routine, such as a street sign, the way we park or even our table utensils. For it's often these things we take for granted and hardly notice – what he called the 'infra-ordinary' – that have the most meaning in our lives. As Moran concludes, 'the smallest details of mundane life can tell us stories about much larger national and global changes.'[7] As we'll see, it's the same with greetings.

2

What's in a Greeting?

'm standing at the arrivals hall in Heathrow's Terminal 5. A flight has just arrived from Madrid and the passengers are beginning to emerge from the double doors, looking crumpled and weary. Next to me a little girl jogs on the spot, holding a bunch of flowers. Her dad is on tiptoes scanning the crowd. 'Mummmy!' shouts the girl, as she ducks under the barrier and runs to meet her. She's swept off her feet and wraps herself tightly around her mum. The dad is more restrained and moves along the barrier. 'Hi, love – you all right?' he says, before going in for a kiss and a hug. Similar scenes light up all around. A couple of lads approach each other, arms wide open, before giving each other meaty slaps to the back. 'Hey bro, how's it going?' Then there's a shriek from behind as a guy in skinny jeans charges through the crowd to meet his partner. These are moments of complete joy and, I'm guessing, some relief – after all, wherever they've come from, the arriving passengers have just been hanging in the air at 30,000 feet. Behind me it's a slightly different affair. A row of smartly dressed men are holding signs with the names of their clients. A passenger points to one and walks over. 'Morning, sir. How are you?' 'Fine, thanks,' he says, shaking his hand.

I've been here all morning watching these little displays. It feels a bit strange, since I'm not actually meeting anyone myself. Instead, I'm armed with a notebook and pen. It's my first day on the job and I'm hoping to work out what exactly is in a greeting.

Despite getting some funny looks, I'm telling myself it's what I have to do and that I'm following in the footsteps of the great ethologists – people who spend their time watching animals in their natural environments in order to spot patterns and make sense of their behaviour. Surrounded by steel columns, banks of glass and Costa cafes, Terminal 5 feels about as far away from nature as you can get, but in many ways it's the perfect place to watch humans. As the philosopher Alain de Botton reflected after a week as writer-in-residence at Heathrow, if a visiting Martian asked to be taken to a single place that captured life in the modern world, an airport would be hard to beat.[1] And the arrivals hall is the ideal spot to observe greetings. In the last couple of hours, flights have come in from everywhere, from Manchester to Mumbai. This really is humanity on full and unreconstructed display.

So, to start at the beginning, what exactly is a greeting? The *Oxford English Dictionary* – the starting point for so many studies – gives the following definition: 'a polite word or sign of welcome or recognition; the act of giving a sign of welcome; a formal expression of goodwill, said on meeting or in a written message'. Or there's Lucy's explanation when she finds herself in Narnia and meets Mr Tumnus, the faun, who gives her a confused look as she introduces herself and holds out her hand: 'People do it... when they meet each other.' In short, our greetings are little routines which we learn and do out of politeness or habit. Yet, although all this might capture the spirit of greetings, and I'm hesitant to challenge the 'world's most trusted dictionary', something here is missing – something more fundamental that might better explain Lucy's (and most people's) thinking.

I turned, then, to what the academics had to say. As I discovered, a leading light, and someone we'll keep coming back to, has been the Canadian American sociologist Erving Goffman, who, working between the 1950s and the 1970s, was one of the most influential thinkers in his field. Unlike most of his peers, who were trying to make sense of the overarching structures and socio-economic trends that shape society, Goffman turned his attention to much smaller, everyday matters. Observing that most people spend most of their lives surrounded by other people, whether in groups and gatherings or among strangers, he set out to identify the various patterns and rules that govern our day-to-day conduct and social interactions. To this end, he zoomed in on the sorts of behaviour that most of us tend to take for granted, such as a passing conversation, ordering in a restaurant or buying something in a shop.[2] Whatever the grand theories, for Goffman, it was in these small-scale, face-to-face interactions that society began.

Famously, Goffman even examined the kind of half-exchanges that characterise many of our interactions with strangers, such as a fleeting glance or moving out of someone's way on the street. We may not give them much thought, but it's these small acts that signal our respect for other people's personal space and the fact that we don't mean any harm. They're what make city living and travelling on the Tube bearable. Goffman coined the term 'civil inattention' to describe this sort of unfocused interaction.[3] In one of his short stories, the author E. M. Davey captures what he was getting at: 'When our eyes met I did that funny thing which is not quite a nod of greeting, nor even the bat of an eyelid, but something subtler: the tiniest flicker of recognition that another human being is in our presence.'[4] While Goffman didn't use the term himself, he's been widely regarded as the pioneer of 'microsociology'. If we imagine that society is a giant termite mound, then the microsociologist focuses on the activity of the individual termites to understand how the overall structure holds together.

Goffman's key insight here is on the importance of 'ritual'. While we tend to think of the term in a lofty way, associating it with mysterious tribal practices and religious ceremony, often an unfortunate goat, Goffman took a wider and more grounded view. For him, rituals were simply those routines and patterns of behaviour that bring people together, and he saw that our everyday lives are full of them. Everything from sitting down to eat or going for an after-work pint to playing a game or watching telly – they're all based on what Goffman called 'interaction rituals'. It's not so much that the activities are important in themselves, but that they bring about joint focus and attention. They are symbols of something bigger. At a more mundane level, Goffman included all of the little unwritten codes and practices that govern our day-to-day encounters and make our public lives manageable, such as queuing in a shop or letting people off a train.[5] In short, from the remotest tribes to inner cities, rituals are the key to social order.

Goffman showed how our greetings are a vital element in all this. Essentially, these patterns of behaviour, whether an elaborate handshake or simple 'Hi', open our interactions, marking the transition from a distant state of civil inattention to focused communication. We use them to negotiate and incorporate ourselves into a social setting. They're what he called 'access rituals' or, along with goodbyes, the 'ritual brackets' that frame our encounters.[6] Without greetings, our interactions would become unmanageable, like a train set without any connectors.

Yet even though Goffman's analysis helps us to see the vital function of greetings, standing in Heathrow, watching the bursts of emotion and even the more sober exchanges behind, I couldn't help feel that he'd missed something. Given how elaborate and intimate these rituals can be, surely they must have some meaning beyond managing our interactions. Here we are helped by the American sociologist Randall Collins, based at the University of Pennsylvania, who, following in the footsteps of Goffman, describes himself as a 'radical microsociologist'. Taking Goffman's notion of interaction rituals, Collins injects them with extra life and meaning. For him, what's most important is not so much that they maintain social order but that, by bringing about our joint focus, they create group consciousness and solidarity. The most successful rituals trigger a heightened state of physiological arousal, leading to a feeling of 'emotional entrainment'. In other words, it's through our rituals that we become one.[7] It's why so many involve a high degree of physicality, in which we try to synchronise our bodies and minds. Think of how many rituals revolve around song and dance – think of the conga. It's these moments of intense energy and emotion that mark the high points in our lives, both as individuals and as social animals.

All this goes back to the founding father of sociology himself, the French intellectual Émile Durkheim. Working at the end of the nineteenth century, in the wake of the Industrial Revolution, Durkheim was concerned about what bound society together, particularly when so many traditional norms and standards were being undermined by such huge social change. So he turned his attention to what was still the biggest bonder of all: religion. From small-scale tribal societies to the major global faiths, Durkheim observed that, beyond the core beliefs, religious life involved regular ceremonial and ritual activities that brought people together and created a sense of solidarity – a 'collective effervescence', as he called it. It was in ritual, then, that he found the key to culture and society. Today, you only have to go to a football match to find an almost religious fervour. Beyond the players themselves, chasing the round symbol of their collective efforts, the crowds combine and channel their emotional energy, chanting songs or doing a Mexican wave. You can get a much better view on TV, but everyone would rather be at the game cheering together.

By connecting Goffman's investigation into public order and Durkheim's concerns about social solidarity, Collins helps us to appreciate the full significance of our interaction rituals – including our greetings. It's through these mini ceremonies that we mark our coming together, both symbolically and literally. With our physical and vocal gestures, we transmit our emotions and feelings towards each other, reaffirming relationships or starting new ones. This was what was going on at Heathrow – greetings as 'intimacy rituals', not just access rituals that regulate our encounters.[8]

But my favourite way of looking at greetings comes from the German sociologist Tilman Allert, who's currently based at the University of Frankfurt. Also following Goffman's lead, Allert is best known for his analysis of what's possibly the most notorious greeting of all: the Nazi salute. In setting out the broader context of his study, he characterises greetings in an unlikely way – as a 'gift' – suggesting that we are in fact offering the greatest gift of all: ourselves.[9] This characterisation of greetings as a gift captures the spirit and importance of exchange and draws on the work of the celebrated French anthropologist, and one of Durkheim's students, Marcel Mauss.[10] Based on his study of

Native American and Polynesian societies, where gift exchange was an important means of developing and maintaining social ties, Mauss's key finding was about the importance of reciprocity – that is, giving back was essential. In Polynesia, for example, if someone failed to reciprocate, whether in kind or with some social favour, they were considered to have forfeited their wealth and authority. While we might like to think that our presents express our affection or appreciation rather than building up or repaying our debts, recent studies suggest that, when we give, we often expect something back or are paying off a social favour.[II] So Mauss gives intellectual rigour to the old adage that 'there's no such thing as a free lunch'.

But whether Mauss's view is always right (my nieces and nephew might need to rethink all the birthday and Christmas presents I have given them), it's fair to say that the expectation of return is no more absolute than with a greeting. If you are in any doubt, try not returning one or think about any time someone has not greeted you back, even if unintentionally. It's no surprise that some of the most famous greetings are greetings that haven't been returned. Consider, for example, the drama around the first time the Premier League footballers and former teammates John Terry and Wayne Bridge met on the pitch after it had been revealed that Terry had had an affair with Bridge's former partner. As the two teams lined up for the pre-match handshake, with over 40,000 fans and the world's media watching, the question on everyone's minds was 'Will they or won't they?' As Bridge came level, Terry held his hand out, but Bridge pulled his arm away, fixing his opponent with an icy stare. Bridge's team went on to win the game, but the next day it was the 'non-handshake' that all the papers were talking about. While less personal, one of the most symbolic snubs took place at the 1954 Geneva Conference. With the participants hoping to bring an end to the Korean War, there was a particularly tense moment when US secretary of state John Foster Dulles came face to face with Chinese prime minister Zhou Enlai. Surrounded by onlookers, Zhou offered his hand, but Dulles apparently brushed past, signalling America's hostility. For Zhou and the Chinese, concerned about saving face, the incident became legendary, poisoning relations with the United States for the next two decades.[12]

Ultimately, it's the New Zealand anthropologist Raymond Firth who gets to the heart of what makes greetings – and therefore any snubs – so significant. Stripped to its basics, a greeting is, in Firth's words, 'the recognition of an encounter with another person as socially acceptable'.[13] This explains why a rebuff hurts so much – it invalidates the essence of what it means to live together as humans. Just as a greeting might be the ultimate gift, not returning one is the ultimate rejection.

So it's clear that our greetings are both functional and symbolic, managing our interactions and expressing our relationships. But, most clearly of all, it's in their absence that we see their true meaning.

———

While it's been useful to step back and consider greetings as a whole, another way at them is to get up close and look at what's inside. We tend to think of a greeting as a single action but, like hitting a tennis or golf ball, there are a number of elements that need to be executed to pull the move off. It was one of the pioneers of the study of face-to-face interaction, Adam Kendon, who, in the late 1960s, first analysed greetings as a sequence of behaviour. Working in New York, Kendon carried out his research at a birthday party and a wedding, employing camera crews at each venue so he could watch the greetings in slow motion, capturing every detail. Based on his observation of seventy greetings in all, he showed how they have a definite order, with a beginning, an end and a number of distinct phases and elements in between.[14] Nearly half a century later, Kendon's study remains a classic of its kind and I'd come to think of him as the godfather of greetings.

Now in his eighties, Kendon still teaches in Cambridge, which is where he began his studies. Having contacted him about my own project, I was invited to visit him at his house; he said that he could even show me some of his original footage, which, for a greetings geek, was a strangely exciting prospect. As I rang his doorbell, I was curious to see how the world's authority on greetings would greet me, though I was also wondering whether I should call him Adam or Professor Kendon. But before I got the chance to say anything, he pulled the door open and then turned away without so much as a hello. I don't think it counted as a snub, but it was a disconcerting start. He began to warm up as I admired the series of cat photos on his wall.

We sat in his living room-cum-library, and I asked how he came to study greetings. 'I started out as a zoologist, looking at snails and birds,' he said. With his white beard and wise eyes, there was definitely something of Charles Darwin about him. He explained that he had gradually become more interested in human behaviour and was inspired by Goffman's early work. In the early 1960s, the social sciences were still new to Britain, so Kendon took himself to the United States, where he connected with Albert Scheflen, a professor of psychiatry based at

Bronx State Hospital in New York and renowned for his work on the management of interpersonal relations. Using video recordings, Scheflen showed how psychotherapy sessions had a common structure: a clear beginning, middle and end, which were marked by particular body movements. It was while building on Scheflen's approach that Kendon came to greetings as a subject of study. Not only were they 'complete interactional events' but, crucially, they were short enough to collect and examine easily.

In the vein of Goffman, Kendon explained that the essential point of greetings is to negotiate a basic social problem: how to move into each other's presence and initiate interaction. Each time I talked about a greeting in terms of a handshake, kiss or some other gesture, he reminded me that they were only part of a wider process. I found myself suggesting that this functional view helped to explain how he greeted me. 'I think I just opened the door and said come in,' he recalled. 'I knew who you were and why you were coming, so it didn't require anything more.' He then thought for a moment. 'I suppose I'm an Englishman of the old school – not very demonstrative,' he smiled.

Kendon moved us to his desk, where he pulled up his footage from the birthday party. It was a summer's afternoon in 1969 and the party was held in a large garden with a private beach, to celebrate the birthday of the host's five-year-old son. The boy's mum, dressed in bright dungarees (it was the 1960s), sat in a deckchair waiting for the guests to turn up, with three camera crews positioned to catch all the action. As the first family arrived, she jumped up and headed over, holding her arms open. 'Hi, how are you?' she said, giving the parents a hug and kiss and their little boy a pat on the head. 'Now, here's a greeting between two people who don't like each other,' Kendon said, as another guest arrived. 'You see: no embrace.' It was like watching an early Woody Allen film.

Kendon was keen to show me a scene on the beach with some of the parents standing around chatting. 'Now look carefully at blue shirt,' he said, pointing to a man who was standing by himself. There was a group next to him in conversation. One of them, in a striped shirt, was eating a piece of birthday cake. 'Now watch as striped-shirt takes a bite,' Kendon said. I leaned in, feeling like I was watching elephants at a watering hole. 'See how blue-shirt notices that striped-shirt is sufficiently disengaged for him to intervene and so makes his approach.' He clicked his mouse and rewound the clip. It was riveting stuff. Kendon was clearly in his element as we went through more greetings, pointing out the tiniest movements that reveal how people negotiate their interactions. In the end, it seems that we're all quite predictable. He looked up as something else caught his eye. 'Ah yes, there's the cat.' We both watched as a large tabby crossed the lawn.

It had taken Kendon days of watching the footage again and again – all 4,800 feet of it – before he could make any sense of it. He recalled how in a single moment he suddenly worked it all out, seeing a clear structure that was common to all our greetings, one that fitted his theories about social interaction. It's with Kendon's help, then, that I was able to get my eye in at Heathrow and we can trace what exactly is in a greeting.

First, and often before we've even seen the person we're going to greet, comes the 'inconvenience display'.[15] Essentially, this is the extra effort we make to get to the first practical point where we can meet someone. At Terminal 5, it's the fact that people have come all the way to the airport, maybe even taking the morning off work and using up a tank of petrol. In our family, you generally have to be away for at least six months to get this sort of treatment. Among world leaders, the issue of who goes to the airport to meet any visitors is a serious matter. After all, taking a morning off work assumes different proportions when you're running a country. When Pope Francis visited the United States in September 2015, for example, President Obama came to the runway to greet him, which was seen as a huge honour. Conversely, there have been some notable absences. From my own studies, there was the time when President Nixon travelled to Beijing in February 1972 to open relations with Communist China. Following twenty-two years of diplomatic estrangement, he'd hoped that Chairman Mao would be there to meet him. But, in the event, Mao stayed away, sending his deputy, Zhou Enlai, and a few guards instead. Nixon took the opportunity to make up for Dulles's earlier snub, ensuring that he was the first to offer his hand, but, all in all, it was a distinctly understated welcome, at least compared to how the Chinese had received other leaders. Mao's absence was noted by the watching media and has been a source of speculation among historians. Maybe he just wasn't feeling well, but it's also been suggested that the Chinese, acutely aware of the power of public image, were showing that they were in charge, or at least could not be easily courted.[16]

At the other end of the spectrum are the small movements we make, such as standing up when a friend enters a bar or walking to the door to meet someone when they arrive. Though less burdensome, these remain deeply symbolic. The basic point is that the level of effort we make relates to the nature of the relationship and status of the individual. These inconvenience displays can reoccur throughout a greeting but the principle remains the same – if you want an idea of what someone thinks of you, see how far they come to say hello.

Next comes the sighting and orientation. To greet someone, we need to identify them and signal our intentions; otherwise, we're liable to experience that cringing moment when we've smiled and waved to someone, only to realise it's not who we thought it was. I learned the lesson at a friend's swimming party, when I decided to greet the birthday boy by dunking his head underwater. It was only after I finally let him up for air that I realised that it wasn't my friend at all and I was at the wrong end of the pool.

To avoid this sort of embarrassment, we generally wave or call out the person's name to get their attention. As Kendon explained, the key point is that we try to make eye contact. If they meet our gaze, it's a signal that they're willing to engage and the greetings exchange can begin. Initially, this is marked by a set of actions known as the 'distant display'. The precise order may vary, but invariably it starts with something that most of us aren't even aware of: an 'eyebrow flash'. Generally lasting for around a fifth of a second, it consists of raising and lowering the eyebrows when we first make eye contact. Through his slow-motion analysis of different people from around the world, the acclaimed Austrian ethologist Irenäus Eibl-Eibesfeldt (his last name is apparently pronounced as 'eyeball-I-best-felt') showed that, while it can be hard to see in real time, this fleeting gesture is a universal signal of wanting to say hello.[17] Drawing on the gift analogy, then, the eyebrow flash is our initial offer, even though we might not know it. If the other person accepts, they will flash back. Handily, as dating experts tell us, it's the first sure sign that someone's interested, though without slow-motion equipment it can easily be missed.

Close behind the eyebrow flash comes the smile. There has been considerable debate about the smile, which we will come to later but, for now, the pulling back and upward curve of our lips, often revealing our teeth, can be regarded as a universal sign of happiness. While it was a common feature of all the reunions I watched at Heathrow, it seems that Kendon's party-goers were slightly less pleased to see each other, with only 82 per cent breaking into a smile; but, whatever the variations, we invariably rely on the smile to initiate our greetings and work out if the other person is wanting to engage.

Along with the eyebrow flash and smile, Kendon identified another common movement on first contact: the 'headtoss'. Like the eyebrow flash, it's often hard to spot, involving a quick up-and-down movement. But just imagine for a moment that you're walking down the street and spot someone you know. Smile over to them and notice what happens to your head. Interestingly, Kendon found that the headtoss is often returned with a 'head lower'.[18] And, closely related to the head lower, there's also the nod, which is a quicker action. Kendon sug-gests that these are used to mark the different phases of our interactions.

Among the crowds at Heathrow, some of these head and facial movements were easy to miss. A more visible part of the distant display, though, is the gestures we make with our hands. Most common, of course, is the wave – an extension of the arm, usually presenting the palm. Often, waves are used to get people's attention, but they can also be symbolic, making up the greeting itself. It's widely suggested that the wave originated as a way of showing that we weren't armed. Although a simple movement, there are several variations. First and foremost, there's the 'lateral wave', which involves moving the hand from side to side, usually at the wrist or elbow, or, if we're stuck on a desert island, at the shoulder, with varying speeds and intensity.

Less common is the 'vertical wave', which consists of moving the hand up and down, as if hesitating about whether to ask a question. There is also the 'diagonal wave', or hail, which involves stretching the arm out at 45 degrees and an extra movement of the hand to avoid looking like a Nazi salute.

Perhaps the most famous wave, though, is one that hardly seems like a wave at all: the royal wave. While other monarchs have adopted the gesture, it's Queen Elizabeth who, now in her nineties, still sets the standard. Consisting of loosely curled fingers and a gentle twisting action, it can seem half-hearted, uninterested even. Some interpret this as the Queen's way of signalling her status above her subjects, though this might be reading too much into it. It seems that the real point, for someone who waves so much, is to save energy and prevent injury. If this sounds far-fetched, consider the experience of the Queen's grandmother, Queen Mary. She started out as a vigorous waver, but developed repetitive strain injury. It was only after consulting her doctor that she resorted to the gentler wave that we see today.

As well as a physical gesture, the distant display may involve a vocal exchange, often consisting of a single-word interjection, typically some variation of 'hello', which itself is thought to come from the Old High German term *halâ* or *holâ*, used for hailing people, especially ferrymen. It only came into popular usage as a greeting with the invention of the telephone in the nineteenth century, when Alexander Graham Bell's American rival Thomas Edison suggested it as an alternative to 'ahoy' as a way to answer. In the English-speaking world, common variations include 'hi', 'hey', 'yo' or, if you're from the Nottingham area, 'ay up', which comes from the Norse term for 'watch out', while the global variants are endless. Like the wave, these interjections can be used either to get someone's attention or as the greeting itself, announcing our existence or arrival; but once greeted in this way, we are obliged to reciprocate.

The American sociologist Harvey Sacks found that these little exchanges have a unique status among our interactions. Sacks became interested in the structure of conversation during his time working as a counsellor on a suicide hotline in Los Angeles during the 1960s, using recordings of the calls to conduct detailed analysis. He showed how even the most rambling chat conforms to various patterns and rules. One of his essential findings was that talk tends to come in responsive pairs, whereby the first part demands and, at the same time, partly defines a response. Examples include questions and answers, offers and acceptance, complements and acknowledgements – and, of course, greetings.[19]

The important point about these initial 'hellos' is that, in our conversations, they have the highest priority. That is, they *have* to go first – saying 'hello' at any other point just doesn't work. Going back to Goffman's notion of 'access rituals', this might seem obvious, but Sacks also showed how these interjections constitute what he called 'minimal proper conversation'.[20] In other words, they are the smallest thing we can say to someone without giving or taking offence. This is true whether we're conversing with a stranger or with close family. And this is the beauty of saying hello when abroad. Armed with one word, you can have a complete conversation with anyone. We can even convey extra meaning just by altering our tone. Depending on how we say it, a simple 'hi' can mean 'I'm so pleased to see you', 'Oh my God, I really didn't want to bump into you' or 'Leave me alone – I don't want to talk to anyone right now'.

So that completes the distant display. Whether a smile, wave, 'hi' or even eyebrow flash, it is the smallest unit of behaviour that we can get away with without offending anyone or feeling like something's up. Going about our daily business, they frequently make up our entire exchange. They are greetings on the go. Often, though, they are merely the starter.

The point at which a distant display moves on to something fuller can be unclear, bringing the potential for mix-up and awkwardness. At Heathrow, there was no doubt that the initial waves and smiles were just the build-up. But if we're in a rush or can't face talking to someone, we can slide straight into a good-bye, so that the 'ritual brackets' make up the entire encounter: 'Hi... bye.' Once either party deviates from their path and initiates an approach, an offer of interaction has been made and it becomes hard to avoid without causing offence. In most cases, the approach involves an extra inconvenience display, which might or might not be mutual. People will often quicken their pace or, at Heathrow, even break into a full, Hollywood-style sprint. Highlighting its symbolic importance, the approach is sometimes laid out with a red carpet for dignitaries and super-stars, a tradition that has its origins in Greek mythology. The earliest known reference is in a play from 458 BC by Aeschylus, which depicts the Trojan War hero Agamemnon returning from battle to find a 'crimson path' (the colour of the gods) laid out for him by his wife. Notwithstanding the fact that it was a set-up (Agamemnon's wife was waiting at the other end to murder him), the practice has caught on in more recent times, and is used across the world to welcome VIPs.

Beyond the walking bit, some extra movements are common to the approach. First, once people have made their move, they tend to divert their gaze, looking at the floor or to one side. This is often followed by what Kendon calls the 'body cross', whereby people bring their arms across their chest or touch their neck. Finally, Kendon found that there was a notable increase in 'grooming behaviour', such as stroking hair or straightening a jacket.[21] At Heathrow, this was particularly evident among the drivers, who would adjust their ties or pull up their trousers as soon as their clients appeared.

Essentially, all of this is building to the climax of the greeting ritual: the 'close' or 'contact display'. Typically, this is what we have in mind when we think of our greetings and, without doubt, it's the stage that leads to the most confusion and torment. Generally, when we're around five feet away, we resume eye contact before going in for the final move. The most common and widespread, of course, is the handshake. From remote tribes to international businesses, it's become the universal sign of saying hello – the 'McGreeting', if you like. We will come back to its history later, but it's widely believed to have originated as a symbol of peace – like the wave, it was a way for people to show that they were unarmed and meant no harm. While the right-handed clasp is most common, there are dozens of variations, from the diagonal arm wrestle and finger snap to the fist bump and high five. It's rumoured that the Freemasons use up to twelve different handshakes, altering the position of their thumbs and forefingers, though they publicly deny using any special variations. Following the handshake in popularity are the hug and the kiss, which also vary widely. One recent survey counted forty-five different kinds of hug alone, from the bear hug to the 'clinger'. Again, we will come back to their origins but, across the world, they are used as a symbol of affection and respect.

Standing at Heathrow, I was struck by just how much the handshake, hug and kiss dominated. Of the 108 greetings

I observed that day, there was only one exception – when a flight came in from Mumbai and a young guy bent down to touch his grandmother's foot. Nevertheless, despite the dominance of the Big Three, there are countless other rituals. Eibl-Eibesfeldt suggested that if we were to record every greeting, the list would fill an entire book.[22] Some of the earliest and most complete efforts to do just this were made by the Victorian anthropologists Edward Tylor and Henry Ling Roth. Working at a time when anthropology was still in its infancy, they very much took the armchair approach, collecting the accounts of various explorers, missionaries and traders. In a paper presented to the Anthropological Institute of Great Britain and Ireland in 1890, Roth detailed over 150, which, as well as the more well-known ones, included the likes of thumb-pressing, finger-snapping, hand-clapping, hand-blowing, body-smearing, stomach-patting, rib-drumming, breast-slapping, body-stroking, nostril-squeezing, cheek-sniffing, paddle-waving, Siamese squatting and joy-weeping.[23] There was a whole section on uncovering. In his journals recounting his Pacific voyage, for example, Captain Cook recorded how, on meeting a young women in Tahiti, she 'expose'd herself intirely naked from the waist downwards' before spinning around.[24] Of all the rituals compiled by Roth, the most extreme was another from Tahiti: slashing the body with sharks' teeth and wailing. According to the account Roth used, this painful practice was the 'singular method of receiving an old friend, or testifying gladness at his arrival'.[25]

For now, a few general points can be made. First, and most obviously, many of these displays involve some form of bodily contact, which can be quite intimate. In these initial moments of interaction, we're often breaking physical barriers, embracing and kissing people we've never met before or wouldn't normally want to touch – our bosses or in-laws, for instance. Moreover, they often involve synchronisation and rhythmic harmony, underlining the state of mutual focus and emotional connection. But although we might become momentarily intimate, the touching is largely symbolic. In other words, unless it's our partner, we are not meant to derive any sensory pleasure from our greetings.

Second, beyond the standard patterns, each routine can vary in its execution. As with a stereo equaliser, we can adjust the balance and intensity. Take the handshake, for example. Most obviously, we can vary the pressure, but we can also alter the rhythm and distance. The same with a hug. Whereas hugging my cousin, to quote another cousin, is like hugging an ironing board, my uncle usually squeezes the life out of me. Third, and relatedly, we can employ a range of extras that, to keep the metaphor going, might be called 'greetings amplifiers'. For example, when shaking hands, we can bring in the other hand to pat the arm or, the politicians' favourite, to cover the handshake. Last, but by no means least, there is the question of when to let go. Ideally, there is a mutual release point, but we've all experienced a handshake or hug that went on longer than expected. Kendon showed me a clip from the party in which the host was greeting a female friend. 'See how he drags his hand down her back. It's a little bit prolonged,' he chuckled, before pressing rewind so we could watch it again.

Partly as a result of all of these intricacies, there can be another stage to our greetings, overlooked by Kendon, which could be called the 'corrective' or 'embarrassment display'. These are all the moves, counter-moves and adjustments we make when our greetings go wrong – when there's been a mismatch in the number of kisses or type of handshake, and we try to resynchronise. Generally, these efforts are accompanied by much nervous laughter, apologies and blushing. In short, it's the stuff that the British have turned into an art.

There is, however, one solution to all this confusion, and that's to avoid any physical display altogether. Indeed, there are some cultures in which public touching is positively, even forcefully, discouraged. But even where a handshake or hug might be the norm, some people still opt out, as Kendon demonstrated with me. At Heathrow, where emotions were running high, most greetings were characterised by a high degree of contact. But I did notice that one older British man kept his distance, standing like a guard, as the women he was with exchanged hugs and kisses. Yet, even where there is no direct contact, there is still a physical dimension to our greetings. Like the man at Heathrow, people often tense their bodies or alter their posture. And avoiding contact depends on a mutual stopping point.

The other key part of the close display is vocalisation – both what we say and how we say it. As well as the various interjections or ways of simply saying hello, we often add a specific reference to the person we're greeting. This can be another element that's loaded with doubt, especially if we've forgotten their name. It makes me think of many embarrassing moments, but one in particular when I threw a house party as a student in the United States. It was all going well until I introduced a Chinese girl who'd just moved into the flat below to all of my classmates. I was struggling to remember her name, but found myself introducing her as Lin Biao. She nearly fell over and my classmates looked at me in disbelief. I was unsure what I'd done, but was then reminded that Lin Biao was one of the main commanders during China's communist revolution, and someone we'd been studying in our Cold War class.

In general, if we don't know or can't remember someone's name we fall back on a generic term such as 'mate', 'pal', 'buddy', 'love' or 'dear'. My personal favourite is 'duck', which I used to get when visiting my grandparents in the Midlands: 'Ay up me duck,' the sweet-shop lady would always say. I'd always thought it was because people were a bit mad up there, but it turns out that it probably comes from the Anglo-Saxon word *ducas*, which was used as a term of respect. If we're meeting someone for the first time or who is more senior, we tend to use a formal address or 'honorific' such as 'Mr', 'Mrs', 'Dr', 'sir' or 'madam'. In some languages, the potential for confusion is magnified by the distinction between formal and informal versions of 'you', such as *tu* and *vous* in France or *du* and *Sie* in German. In English, there used to be 'thou' and 'ye', which Shakespeare used to dramatic effect, but this distinction disappeared in the late sixteenth century, at least removing one dilemma. The key point explaining why getting them wrong can cause such discomfort is that bound up in all our names and titles are matters of personal identity and social status.

There are also what Firth calls 'affirmations', which add an element of good wishes, most closely reflecting the dictionary definition of greetings. Often, they refer to a particular part of the day ('Good morning/afternoon/evening'). Or there are the various ways of saying 'Pleased to meet you'. And, finally, there is what could be called the well-being enquiry or 'How are you?' display. Again, there are countless variations. In the English-speaking world alone there are numerous ways to put the question. Beyond the standard phrases ('How's it going?', 'All right?', 'How do you do?'), in the United States there's 'What's up?' (or 'Sup wid it?'), 'Howdy?' (from 'How do you do?')

and, my personal favourites, 'How's it hanging?' and 'What's shakin' bacon?'

Most interesting, and the cause of some debate in academic circles, are the responses we give to these questions. Surprisingly, our replies are nearly always the same, or at least fall within a set range. However we're feeling and whatever version of the question, we generally reply in positive terms. Under the grey skies of the UK, we're generally 'Not bad' or 'Fine', unless we've just got married or won the lottery, in which case we might stretch to 'Pretty good'. The basic point, then, is that whatever the question, it's not an invitation to go on about your bad toe or how much you hate your boss. In short, the question marks are misleading. All of this has led academics to conclude that these exchanges have no informational or propositional content – in other words, they are empty. They are classic examples of what the Polish anthropologist Bronisław Malinowski called 'phatic communion', more commonly known as 'small talk'. It's conversation for the sake of it – the equivalent of keeping the car engine running without actually going anywhere. What's important isn't the content but keeping the conversation going, avoiding any awkward silences. As Collins observes, our capacity for small talk, or sociable conversation, is the most basic of all interaction rituals. To quote the Irish poet William Butler Yeats, these are 'songs written for the tune's sake'.[26]

So, as a matter of course, we often distort the truth. While conducting research in a local hospital, Sacks found that even when patients looked like they were about to die – and some were – they would still respond to 'How are you?' with something like 'Fine, thanks'. The philosopher Immanuel Kant argued that when we lie we violate our self-respect, the dignity of the person we lie to and our duty to humanity as a whole. That may be true, but there are also 'white lies', or what Goffman might have called ritual or functional lies – lies that help to manage our interactions. Unless we're being chased by an axe murderer or someone's died, the real 'How are you?' and honest response come after the greetings. This phatic communication dominates our opening exchanges. Often, it extends into the first topic of conversation: 'How was the flight?', 'Did you get a meal?', 'How was the journey to the airport?', 'Was the traffic OK?', and so on. The truth is that, unless the emergency services were involved, we don't really care.

Before we leave all this phatic talk, one topic should be mentioned: the weather. The British, in particular, are renowned for their weather talk – it's the default starting point for most exchanges, even a greeting itself: 'Nice day, isn't it?' for example, or, more likely, 'I can't believe it's still raining.' People have wondered where this obsession comes from. After all, as Bill Bryson quipped, for any newcomer, 'the most striking thing about English weather is that there is not very much of it.'[27] Bryson touched a national nerve, provoking our most formidable inquisitor, Jeremy Paxman, to hit back. Bryson had missed the point, Paxman argued. We might not get the violent extremes, but the reason we spend so much time gassing on about the weather is that it's so unpredictable. To our endless frustration, from one minute to the next, we don't know whether it's going to rain or shine.[28] It's why watching the weather forecast (and commenting on the presenter) has become a nightly ritual in itself.[29] All this might be true, but when viewed against the notion of phatic conversation and its key social function, we can see another reason for all our weather talk: as the anthropologist Kate Fox puts it, it's the 'facilitator of social interaction'.[30] Like the flight and journey, it's an example of what Sacks calls a 'false first topic', merely smoothing the transition to what, if anything, we really want to talk about.[31] It's what we have in common and, for a nation that struggles with small talk, can be a social saviour. In this sense, 'How are you?' and 'I can't believe it's still raining' are the same. We're just getting our engines warmed up.

If we haven't seen someone for a while, our opening exchanges will often include another kind of conversation that could be characterised as 'bridging talk'. These are comments such as 'It's been so long' or 'When was the last time we saw each other?' Often we'll look each other up and down and pass a compliment: 'You're looking well' or 'You haven't changed at all'. At Heathrow, I noticed a couple of older men inspecting each other: 'You've lost a bit, Tony,' one of them said to the other, as he patted him on the stomach. As well as being a nice thing to say, it's a way of suggesting that we haven't suffered in each other's absence. As we complete our mutual examination, there will often be extra fussing and patting.

Once these physical and vocal elements are out of the way, the greeting ritual is finally over. We leave the special spot on which the ceremony has taken place and resume our normal relationships. At Heathrow, there were a couple of extra elements that were so frequent that they could be regarded as an integral part of the ritual. First, the meeting party would invariably take hold of the passenger's case or trolley – an extra inconvenience display – which would often result in a tussle all the way out of the airport. And, second, bringing greetings into the twenty-first century, people would pose for a selfie, capturing the moment of reunion forever.

There may be one last stage to complete the ritual: the 'gift display'. As if the gift of each other isn't enough, we also present a material offering. If we are going around to someone's house for dinner, this might be some wine or chocolates. In the arrivals hall, the preferred items were flowers and helium balloons, whereas I assume that buried somewhere in the passengers' cases were fridge magnets and 'I love [insert place name]' T-shirts. Among world leaders, the gift exchange is an important part of any visit. Often the gifts will be based on some national emblem – a sort of super fridge magnet. In her sixty years as head of state, the Queen has received some unusual presents, including a giant grasshopper-shaped wine-cooler from the French president Georges Pompidou, a sword made of sharks' teeth when she visited the Pacific island of Kiribati, an elephant from the leader of Cameroon and two sloths from her visit to Brazil in 1968. Thanks to concerns over transparency, we can now also go on the US State Department's website, follow the links to the Gift Protocol Section, and see exactly what the US president gets. Unsurprisingly perhaps, the most lavish givers are the Saudis. In 2014 alone, King Abdullah showered Obama and his family with hundreds of thousands of dollars' worth of gifts, including a jewellery set valued at $80,000 and a gold watch worth $67,000.[32] All of which must have been pretty tantalising, as they're not personally allowed to keep anything over $375. One of the most thoughtful, though, came from Gordon Brown when, on his visit to the White House in 2009, he presented Obama with a first edition set of Martin Gilbert's seven-volume biography of Churchill along with a pen-holder hand-carved from the timbers of the anti-slavery ship HMS *Gannet*. In return, Brown was given a DVD box set of Obama's favourite films, which included the likes of *Star Wars: Episode IV* and *Psycho*. Analysts of Anglo-American relations didn't miss the chance to point out which way the relationship mattered most.

But gifts don't have to be materially valuable. While it's

always nice to get something we actually like, it's the thought that counts. They're symbols of our bonds and the status of our relationships. It's why home-made presents are often the best. Think of when Hillary Clinton, as secretary of state, presented her Russian counterpart, Sergei Lavrov, with a big red button that said 'reset', symbolising a fresh start to relations. It can only have cost a few dollars to make but the sentiment was priceless. The only problem was that the Americans had got the translation wrong, using the word *peregruzka*, which means 'overloaded'. Nevertheless, the thought was there.

3

How to Greet the Queen and the Elusive Art of Etiquette

I t was 11 June 2010 and the Queen was about to visit the Cabinet Office, where I was working. It was a historic occasion, marking her first visit to the department and its role in the negotiations that had just led to the formation of the new coalition government. I'd played absolutely no part in them but, like the rest of my colleagues, was smartly turned out to see her. Someone had been specially selected to present a bunch of flowers but, for the rest of us, there was still a chance that we would get to meet her as she walked down the line. I'd pushed my way to the front to increase my chances. As news spread that her car had arrived, the excitement peaked. If I'm honest, I was also feeling a bit nervous. If she did stop, I still wasn't sure what I should say or do. I'd heard all sorts from various aficionados at work: you should bow, but only with your head; call her ma'am (pronounced like jam), not Your Majesty; whatever you do, though, don't offer your hand. Laughing at all this, my friend John said he was going to give her a high five. There was a hush as she entered the building, dressed in light blue, smaller than expected and looking a bit hesitant. I shuffled forward to make myself as visible as possible. Forget the nerves, this was going to be one for the grandchildren.

Greeting the Queen is about as extreme as it gets. But, whether meeting the head of state or an old friend, the issues are the same. As a result of the last chapter, we might have a better understanding of what's in a greeting and the significance of these rituals, but how do we put all of this together and make sure that we get it right? How do we know which routine to go for? And, even then, the devil is often in the detail: who goes first? What strength and length of handshake or hug to give? How many kisses and which cheek to start with? How to refer to someone and what to do if we've forgotten their name? To refer back to Tilman Allert, we might know that we want to make the offer, but we can still be uncertain about how to make it; and when someone reciprocates the confusion can double. All of which takes us back to my original mission: to see if anything could be done about it.

The best place to begin would seem to be with the etiquette advisors. After all, these are the people who have made a profession out of telling the rest of us how to behave. At home, the most famous source of correct behaviour is Debrett's. Established in 1769 as a publishing house, it started out producing a yearly who's who of high society, though since the early twentieth century it has also offered guides and coaching on all aspects of social life in Britain. While the British claim to be world leaders, other countries have similar organisations. In the United States, the term etiquette has become synonymous with the name Emily Post, whose guide, first published in 1922, is now in its eighteenth edition. Today the fourth generation of her family is at the helm and there is an entire institute under her name. Similarly, in Germany, one person, above all, has come to symbolise proper conduct: Adolph Knigge. While he is still known for his eighteenth-century treatise on human relations, the society that takes his name today has become notorious for its strict approach to social kissing at work. For centuries, then, etiquette advisors have been setting us straight on how we should behave. More recently, however, etiquette has suffered from an image problem. The experts are seen at best as charming relics from a bygone age, and at worst as insufferable snobs, deepening social division. Mostly, though, they are regarded as comic throwbacks who add a bit of novelty value, and their books are generally tucked away in the humour or gift section.

But should we be so quick to dismiss them? It makes me think of the spate of books bemoaning the decline in standards of grammar. Their authors complain that grammar is no longer taught or properly understood. From the other corner, they're dismissed as 'grammar Nazis', stuck behind the times and obsessed with arbitrary rules. Does it really matter whether you split infinitives or know what an Oxford comma is? Language is fluid and always evolving, they point out. This might be true, but I have to admit that one of the books that had an immediate and lasting impact on my life was a small guide to punctuation. By the end, I finally knew how to use a comma, dash and even a semicolon. It was a revelation: rather than being oppressive, these rules were liberating. They could help me express myself and be understood – which, after all, is surely the point of language. Could etiquette guides do something similar, freeing us from uncertainty and mistakes? As one expert promises, they hold 'the key that opens the door of greater social happiness'.[1]

It was as I was driving home one afternoon before Christmas, listening to the radio, when I heard just the person I needed to meet. He was talking in clipped tones, offering all sorts of seasonal advice, from how to eat a mince pie (always with the left hand, so you're free to greet people) to how to decorate your house (with white lights, never tinsel, and keeping cards from anyone important at the back, though clearly visible). This, the audience was reminded, was William Hanson, 'the UK's leading etiquette expert'. He was being treated slightly mockingly, as he used words like 'vulgar' and 'interregnum', but he had answers to every social dilemma. Back home, I found his website, complete with a large photo. Dressed in a blazer and violet tie, with a crop of blonde hair slicked to one side, he certainly looked the part, if some years younger than his voice and profession suggested.

Trawling Hanson's website, I found advice on everything from how to hold a teacup to how much to tip. Some of it made good sense, such as his rules on 'jetiquette': try not to invade fellow passengers' territory, stop children from kicking the seat in front and remember to thank the crew. Some of it less so: floral bedding is only acceptable in thatched houses and women shouldn't wear trousers after six. Most cryptic was his latest tweet: 'You can tell a lot about a man by whether they own a black or a brown wallet.' *Where did all this stuff come from?* Although he clearly had a wide following, it seemed that Hanson's advice wasn't always received positively. In one of his videos, he read the feedback he'd received to his *Daily Mail* column, cheerfully relaying various abuse, from being called a pretentious plumb and misogynist dinosaur to being told to get a proper job. Funniest of all, though, and seemingly Hanson's favourite: 'William Hanson has the kind of face you would never tire of slapping with a haddock.' 'Why a haddock?' he chuckled. Whatever his critics, here was someone who played the part and could laugh at himself.

And so, on a drizzly January morning, I set off for Manchester, where Hanson was based. He'd arranged for us to meet at the Great John Street Hotel, which sounded suitably grand. As my train made its way through the Pennines, I leafed through Hanson's best-selling *Bluffer's Guide to Etiquette*, along with the latest advice from Debrett's and the Emily Post Institute, to see what they had to say on greetings. Focusing on first meetings, overall, the advice could be roughly summarised under four principles. First, it's generally best to err on the side of formality. While you can always relax the tone later, it's much harder to pull back. Second, when working out who goes first or what greeting to go for, you should pay attention to precedence or social hierarchy, which mainly means deferring to women and older people. Third, if in doubt, follow the other person's lead. Finally, and perhaps most importantly, make sure all this seems natural and sincere. As a general rule, affectation is the biggest enemy of etiquette.

In terms of translating all this into the greeting itself, the handshake comes out on top, accompanied with a smile and eye contact. Gloves should always be removed (unless it's freezing outside) and the aim is to grip the other hand firmly enough to 'leave an impression of bodily contact'.[2] Sensibly, men are advised to be gentler with women, to avoid squeezing any big rings. As for the shaking part, the general consensus is two or three pumps, five or six inches in height. Any extras, such as cupping the hands or patting the shoulders (what I called the 'amplifiers') are strongly discouraged. And, except with close friends, all variations are to be avoided. Tweeting a picture of Prince Charles receiving a diagonal clasp, Hanson made his feelings clear on the matter: 'I don't know about you, but this makes me violently ill.'

On the trickier issue of social kissing, special caution is advised. In fact, reading across the guides, it's fair to say that etiquette advisors have been almost as reluctant as me to accept the trend, describing it as an 'epidemic', something to manage rather than embrace. As a rule, it's best to leave out the kisses when meeting someone for the first time. And when you do go in for a peck, keep it cheek to cheek, kissing just the air: 'under no circumstances should there be suggestion of saliva.'[3] As for how many to go for, unless you know someone well, one is considered enough, with the 'double-pronged attack' largely left to the Continentals. However many, you should always start right, tilting the head slightly left to avoid collision. Overall, though, when it comes to kissing, the central message is: 'If in doubt, don't.'

As for what to say, the advice is more divided, sparking some strong opinions. According to the most recent Emily Post guide, some form of 'How are you?' or 'Pleased to meet you' is considered to be a safe start. But the British guides generally advise against these openers, since it's felt that they break the number-one rule of etiquette: sincerity. As Hanson puts it, if you've not met someone before, you don't really care how they are and can't know if you're pleased to meet them. Instead, the preferred greeting is 'How do you do' (there's no question mark) followed by the person's name. The gesture is simply repeated back. If you find yourself in the thorny position of having forgotten someone's name, there's always the old trick of introducing the person whose name you do know and hoping that they oblige. If this doesn't work, you could try Hanson's method of saying 'I'm sorry, I've forgotten your name'. When they reply, you then say 'No, I meant your surname', though, as Hanson warns, this takes some confidence. If you meet them again and still forget, you could try the reverse tactic of saying 'Of course you won't remember my name'. Finally, it can be best to come clean, making the excuse that you have a terrible memory.[4]

While much of this made sense, I have to admit that I didn't experience the same epiphany I'd had when reading my punctuation guide. Many questions were left unanswered and the final fallback, 'if in doubt, follow the other person's lead', seemed a bit of a cop-out. What if both people are in doubt? Moreover, while I understood Hanson's reasoning, I couldn't imagine saying 'How do you do' without sounding insincere (the biggest mistake) or laughing. I started to worry about what I was going to say when I met him. And, reading his section on dress, I was beginning to regret my decision to wear an orange mac and brown loafers. 'Even with jeans or chinos on a Tuesday, wear black shoes... or any other colour than brown,' he instructed. 'Just remember: don't wear brown in town.'[5]

As I entered the hotel – more new boutique than Claridge's – I spotted Hanson in the corner dressed in chinos and a stripy jumper that had a hint of Christmas about it. I was disappointed that he wasn't turned out in a blazer and cravat, but at least felt better about what I was wearing. Then, as he stood up (the 'inconvenience display'), I noticed his shoes: *brown* brogues. What was going on?

'Hello. How lovely to meet you,' Hanson said.

Hang on, weren't you meant to say... This guy was breaking all his rules before I'd even had chance to question them myself.

'What would you like?' he asked, as the waitress came over.

'I'll go for a tea,' I said, really wanting coffee, but assuming it was the done thing.

'Perfect,' Hanson said. At least I'd got one thing right. 'And I'll have a coffee.'

Once we'd got through the small talk about my journey and the dismal weather, Hanson explained that he'd just returned from China and India, where he'd been teaching one of his courses: 'The Secrets of Elegant English Entertaining'. It pre-empted my question about where his main business comes from. Apparently, the biggest markets are now in the Far East, where interest in etiquette is booming among the rising middle classes. Until recently, the clientele had been mostly Chinese businessmen, seeking to understand Western customs and consumers. Now, though, it's mainly wealthy women trying to refine their social skills and prepare their children for an education abroad. In India, the interest in etiquette combines a romanticism about how things were done during colonial times with a new global outlook. For Hanson, the good news in all this is that the British are seen as the leaders. Whatever our diminished status on the world stage, when it comes to etiquette, we're still a superpower. I joked that all this was ironic, given our reputation for being so awkward. But, for Hanson, it touched on a serious point. While business was thriving abroad, back home, people took a different attitude.

'We're British. We know how to behave,' Hanson said sarcastically. 'Actually, a lot of people don't.'

Before we got too far into the state of etiquette at home, I was keen to find out how he had got into his line of work, and what qualified him and others to tell the rest of us how to behave.

'It was an accident,' Hanson admitted. 'Really, I wanted to be a newsreader.' As it happened, his career path began at the age of twelve when his grandmother had bought him the latest Debrett's guide for Christmas. I guess this is how many lives are defined. At a similar age, my grandparents had bought me my first bird book, sparking a lifelong passion. The only difference was that, while I was hunting for sparrowhawks, Hanson was honing his dining skills. By sixteen, he had his first clients. 'By clients,' he said, adding inverted commas with his fingers, 'I mean members of the WI and Boy Scouts.' But Hanson gradually developed a name and notoriety, partly due to his young age.

Now, a decade later, his 'clients' include diplomats, business leaders and royal households, while he travels the world and makes regular appearances on TV.

Our drinks arrived and I was confronted with an enormous pot, which had a long, curved spout that threatened to send tea everywhere. Remembering the instructions from Hanson's book, I poured the tea first and then added some milk. But I misjudged and was left with the kind of drink you'd bottle-feed a baby. I could sense Hanson's eyes watching. *Is it OK to add more tea at this point? Don't be so ridiculous,* I told myself, as I poured more, but now my cup was overflowing and, as I tried to take the first sip, my shaking hand sent it everywhere. Hanson was kind enough to look away, but it underlined one of my concerns about all this etiquette stuff: while trying so hard to do the right thing, it becomes impossible to relax.

I asked Hanson how he decides what's correct behaviour. 'Where do these rules come from?'

'Generally, it's trial and error,' he said. 'We pick the best bits from different times. It's the Victorians and Edwardians, with a bit of the Georgians thrown in.' He paused for a second. 'Mostly, though, it's Victorian values that have stuck.' I'd hoped to get a bit deeper, maybe even philosophical, but it was clear that Hanson was mainly taking his lead from what he saw as a golden age of behaviour.

Before I could work out how to raise the issue, Hanson jumped in: 'You see, it's all about class. In Britain, *everything's* about class. In a way, etiquette is about showing that I'm better than you'. Although I'd warmed to Hanson, I could see what got his detractors going. Seeing me frown (I was caught somewhere between disbelief and recognising that, whatever our politicians say, he had a point), Hanson set out to explain. He referred to an article written by the linguist Alan Ross in 1954, in which he distinguished how the English upper class speak, inventing the terms 'U' (for upper class) and 'non-U' (for the rest) to make the distinction. Aside from the difference in pronunciation, Ross came up with a long list of words and phrases, giving their 'U' and 'non-U' equivalents respectively, such as 'to have one's bath' as opposed to 'take a bath', 'tea' rather than 'cup of tea', 'riding' not 'horse-riding', and saying 'sorry' instead of 'pardon'.[6] In part, Hanson explained, this was the point of etiquette: to distinguish oneself as upper class. It was why it was important to know how to hold a teacup and set a table properly.

I was about to challenge Hanson, but he went on to qualify his view of class, bringing in the author and socialite Nancy Mitford. In her essay on the English aristocracy, in which she popularised the notion of 'U' and 'non-U', Mitford argued that language was the only thing left that distinguished the upper class.[7] It was no longer about inherited wealth and status. To be upper class, you just needed to say and do (and wear) the right things – it was a question of style. In that sense, it was a democratic view of elitism – anyone could be an aristocrat. And so it was with Hanson. After all, as he explained, his grandmother had come from humble origins. But it was her belief that everyone should be able to speak to a duke as much as a dustman. Seen in this light, Hanson's mission wasn't to deepen class divisions but to create one class: the upper class.

Much of this framed our discussion on greetings. As well as putting people at ease, the opening interaction was about showing yourself to be from the right class, Hanson explained. It's why he felt that greetings had become too informal. He rehearsed the guidelines from his book: 'The correct form is to say your first name, then your last name, followed by "How do you do", which is then repeated by the other person.' The crucial point about 'How do you do' was that it was very much 'U'. 'Pleased to meet you,' he said, putting on a Britney Spears accent, 'is most definitely not.' I challenged him on his greeting to me, but he explained that we had exchanged emails already and developed a rapport. 'When I said it's nice to meet you, I meant it.' Again, the guiding principle was sincerity.

As for the physical side, Hanson emphasised the virtues of the handshake: 'In my eyes, a handshake will never be wrong.' I couldn't resist asking how mine was. 'Yes, I did notice – nice and firm,' he said. I felt like I'd passed an important test.

Turning to social kissing, Hanson was more wary. In fact, to my surprise, given how I'd come to view the double kiss, it really wasn't 'U' behaviour at all. 'You know, I've been to parties where girls have wandered in, usually wearing too much make-up and high heels, lunge in and give a social kiss,' he said, looking like he'd swallowed a wasp. 'I keep my arm locked out,' he added, demonstrating the motion. 'That normally stops them.'

'But what if you're on a first date?'

Hanson was adamant. 'People think that we're on a date so we'll kiss. No – you haven't met them before.'

He conceded that you could always warm things up with a shift in tone or bigger smile. 'Or you could stand sideways on to them,' he added. He'd started to lose me.

I pushed Hanson on some trickier situations, such as the follow-up date or meeting parents' friends you haven't seen in years, or a friend's friend. 'In the end, it comes down to judgement,' he said. 'It's hard to teach someone emotional intelligence. All I can do is show people the basic rules and they can then slide up and down accordingly.' He then fell back on the usual advice about following the other person's lead and being confident.

I sensed that I was beginning to squeeze Hanson dry, and he had a Skype call with potential new clients in Saudi Arabia to get back to. But I'd saved up one last question: 'What's in the colour of a wallet?' Hanson's eyes rolled. 'God only knows why I tweeted that. I must have been drunk.' Apparently, it had set off a storm among his followers. He squirmed, as I showed him my brown wallet. He was struggling to explain but, in the end, it came down to the fact that brown was more casual. I could see what he meant, but enjoyed watching him strain.

I left our meeting feeling confused. On the one hand, for all his pomp, he was actually refreshing – an antidote to our politically correct and casual times. Seen in the most favourable light, he also had an important message, mission even. At a time when we're becoming more mixed up than ever, Hanson was promoting common standards that would help us interact, rooted in a basic level of respect and consideration. He also had some useful advice on greetings. On the other hand, he was the biggest snob I'd ever met. However he couched it, his obsession with class seemed to cloud his view of behaviour, which I felt was undermining his cause. The idea that manners are the badge of the upper class is enough for everyone else to reject them. And I wasn't convinced by some of his advice on greetings, either. Even though I sympathised with his views on social kissing, watch any episode of *First Dates* and you'll see how much the double kiss has become the norm. If anything, it's the single kiss that is saved for the people we're closest to. As for 'How do you do', I just couldn't imagine using it. Also, somewhere in all this, I recognised that Hanson was living up to a certain persona. It was just hard to know where to draw the line.

To make the most of my trip to Manchester and get a different perspective, I headed to the People's History Museum, a homage to democracy and working-class Britain. Filled with colourful trade-union banners, workers' manifestos and photos of social activists, the overarching theme is the fight for equality and overthrowing the ruling elite. This was very much the story of the anti-U, a celebration of working men's clubs and cooperatives, where I imagined that any talk of napkins and 'How do you do' would get short shrift. A comment scrawled on the visitors' board set the tone: 'Overthrow the patriarchy.' With Manchester regarded as the first industrial city and cradle of the Labour and suffragette movements, the museum is well located. Walking around the centre afterwards, I wandered into various shops and cafes to see how I was greeted. Invariably, I got some version of 'All right, mate' or 'All right, love'. While these wouldn't have passed the 'U' test, they were warm and welcoming, which was surely the point.

———

Returning home, I was beginning to think that, in trying to find the right way to do things, I'd made a wrong turn. But before I completely dismissed the advice of the etiquette advisors, I wanted to go back to their founding fathers/mothers, to see what the Victorians and Edwardians had to say. As far as possible, it was with an open mind that I sat in Cambridge University Library waiting for my books to arrive. It's an imposing place, one of only five legal deposit libraries in the country, which means that it's entitled to a free copy of every book published in the UK. It also contains such treasures as the Guttenberg Bible and the papers of Isaac Newton. I'd felt a bit sheepish about my own order, a selection of old etiquette books. There was Emily Holt's *The Encyclopaedia of Etiquette*, published in 1901 and considered a classic, F. R. Ings's *Etiquette in Everyday Life*, which had the promising subtitle *A Complete Guide to Correct Conduct for All Occasions*, from 1900, and, most famous, the first edition of Emily Post's *Etiquette*. Influenced by much of what Holt had to say, Post had originally set out to produce a small and simple book, but ended up writing over 600 pages of advice.

Glancing through their browned pages, I felt like I'd been transported to another world, one where the rules were clear and everyone knew their place. It was like reading a scene-setting for *Downton Abbey*, only without any fun. There were stipulations on everything, from directions on how to fold a napkin to a long list of phrases to avoid. Most striking, though, were the instructions on the correct length of mourning: between six weeks and three months for an uncle or aunt, six weeks for a cousin and three weeks for a second cousin, though this is not obligatory. There was no explanation. The important point was to follow the rules, not question them.

In conversation, the aim was to 'try to... say those things only which will be agreeable to others'.[8] As Post warned, 'the man who has been led to believe that he is a brilliant and interesting talker has been led to make himself a rapacious pest.'[9] Above all, though, it was interactions between men and women that generated the strictest guidelines, as if they inhabited separate spheres in life. On coming into contact, the basic rule was that men should show deference and respect to the 'fairer sex'. For example, if a lady was standing, at no point was a man allowed to sit. However constrained and baffling these rules were, the key thing with all of them was to appear effortless and natural, without any hint of affectation. As G. R. M. Devereux put it, 'the perfection of manners is "no manner".'[10]

These broad principles set up the advice on greetings. There was no confusion about who goes first: 'It is the privilege of the lady on all occasions to make the first move.'[11] Above all, though, the guides stressed that they were to be made without fuss. While it was fine to show pleasure at seeing a friend, Ings warned that 'they will not expect you to make a scene'.[12] As for introductions, Post was adamant: 'You must never introduce people to each other in public places unless you are certain beyond doubt that the introduction will be agreeable to both.'[13] In accordance with the rules on deference, it was up to the women to take the lead, though Holt warned that going around offering a hand was a mark of 'impulsive provincialism'.[14]

Handshakes were to be made briefly, but leaving a feeling of strength and warmth. As Post memorably put it: 'Who does not dislike a "boneless" hand extended as though it were a spray of seaweed, or a miniature boiled pudding?'[15] I could hear Hanson chuckling at that one. In contrast with today, though, the most common form of greeting was the bow. The art of bowing was said to be one of the most complex subjects to grasp. No wonder, when this was the sort of advice on offer:

> When meeting a gentleman alone do not raise your hat, simply nod; but if you are with a lady who bows to a gentleman, whether he is known to you or not, raise your hat at once; the same rule applies if the lady you are with bows to the other lady. When meeting an old man, or one your superior by rank, birth, or learning raise your hat, and when introduced to such raise it slightly and bow also slightly.[16]

As for what to say, there was only one safe option: 'How do you do'. 'Hello' was considered vulgar, while it was said that 'persons of position do not say "pleased to meet you"'.[17]

Reading all of this, I could see where Hanson got his inspiration. But it also underlined the fact that times have changed. Even Hanson was making concessions and adding humour in order to get heard. It was like he was grabbing at sand – the tighter he squeezed, the more it drained away. Looking at all these arbitrary rules also made me think back to a claim that Hanson makes at the beginning of his book: that etiquette and manners are the same thing. Digging into the origins and history of the two terms shows that the relationship is not so straightforward, or at least that there are different interpretations. We tend to think of manners as polite behaviour we do out of consideration for others, whether saying hello or opening a door for someone – what the seventeenth-century English philosopher Thomas Hobbes called 'small morals'. In this sense, we might regard them as universal and timeless. Etiquette, on the other hand, is more particular, referring to the codification of our behaviour. The term itself goes back to the Palace of Versailles under Louis XIV, when the head gardener put out little signs or tickets – *étiquettes* – telling people to keep off the grass. The term caught on and gradually came to cover all rules for correct behaviour in royal circles.[18]

But the attempt to codify behaviour through written texts goes back much further. One of the pioneering attempts was by the Dutch humanist Erasmus, who in 1530 published an influential treatise that brought together the various rules and standards that had started to spread through different songs and sayings. His central theme was the importance of 'outward bodily propriety' and, with it, he hoped to shape the next generation, particularly young boys. Bearing in mind that it was still early days, they were instructed to keep their penises out of sight, mask the passing of wind with a cough and avoid greeting someone if they were urinating or defecating. With translations in French and English, Erasmus's ideas spread widely, reflecting what the German sociologist Norbert Elias famously called the 'civilising process' of the Middle Ages, which was characterised by an increase in people's inhibitions and self-control and, accordingly, a lowering of thresholds of shame and embarrassment.[19] It's easy to see the connection to what the Victorian etiquette writers would say about the importance of restraint four centuries later.

While Erasmus hoped to reach a wide audience, the real standard-bearers for manners could be found in the great medieval courts. For the royal courtiers, they were a matter of their profession. To show their deference, they would perform what Randall Collins called 'asymmetrical rituals', scraping to the floor in theatrical bows. With members of the courts living in an almost constant state of ceremony, it's easy to see how they became increasingly self-conscious and prone to embarrassment. As the threshold of shame fell, one of Elizabeth I's favourite courtiers, Edward de Vere, Seventeenth Earl of Oxford, suffered an unfortunate moment when, making a deep bow before the Queen, he farted. Apparently, he was so mortified that he fled the country and didn't come back for seven years.[20]

Although court society set the standards for manners, it also marked their separation from a deeper morality, turning them into a badge of the elite. Within court circles, manners were less about good behaviour for its own sake and more about currying favour and self-advancement. The influential *Book of the Courtier*, written by the Italian diplomat Baldassare Castiglione at around the same time as Erasmus's guide, was all about making the right impression, emphasising the importance of social dexterity and gratification. Manners had become an art – something to craft and hone. Today, we might think of our restaurants and car showrooms, with staff told to put on their best smiles. It's what American sociologist Arlie Hochschild has otherwise called 'emotional labour'.

But it wasn't until 200 years later, in the mid-eighteenth century, that this utilitarian understanding of manners, and the term etiquette, would become popularised in England. More than anyone, the person who has been associated with – and blamed for – this approach to manners is Philip Dormer Stanhope, Fourth Earl of Chesterfield, better known as Lord Chesterfield. A successful politician and diplomat, Chesterfield became most famous for a series of letters that he wrote to his illegitimate son, Philip, whom he worried didn't have the looks or charm to get ahead in the world. However, the central message of his letters was that this shouldn't matter: 'Good breeding,' he assured his son, 'is a mode not a substance.' The key thing was to adapt and imitate those around him. 'Without hesitation, kiss his [the Pope's] slipper, or whatever else the etiquette of that Court requires,' he wrote, bringing the term into popular use.[21]

Chesterfield's advice began to reach a wider audience when his letters were published in 1773 as *Principles of Politeness*. For the socially challenged – people who, like me, struggled with their greetings – they promised an end to self-doubt and awkwardness. Take eighteen-year-old Isaac Mickle, for instance, who in 1841 was having a tough time meeting girls. As he confided in his diary, things came to a head one afternoon when, attempting a bow, he caught his feet in a carpet and fell 'sprawling to the floor'. Since then, he'd hardly managed a nod. Having reached a low point, he resolved to read Chesterfield's *Principles* as soon as possible, anticipating that he would make a new entry into society with 'becoming flourishes'.[22]

Beyond the hope they offered to awkward teenagers, Chesterfield's *Principles* resonated with a wide section of society, coinciding with a period of great social change. Against a backdrop of political and industrial revolution, on both sides of the Atlantic the second half of the eighteenth century saw a burgeoning middle class, newly conscious of the possibilities for getting ahead in the world. For these social climbers, propelled by wealth, as opposed to inherited status, Chesterfield's message of adaptation and imitation would help smooth their rise in society. By popularising manners, Chesterfield had opened the back door to the aristocracy.

But there was another side to all this. For the emerging middle class, the interest in etiquette was not just about gaining access to the social elite but, perhaps more importantly, about distinguishing themselves from the rest. It's no coincidence that the high point for etiquette writing in the Anglo-American world during the nineteenth century was also a time of mass immigration from Ireland and the emancipation of millions of slaves. Devising ever more elaborate and stricter social rules around introductions, proper dress and the like was a way of keeping society segregated. Paving the way for the likes of Holt and Post, manners became a minefield for a reason. While politicians were espousing the ideals of democracy, etiquette guides were doing the unsayable. The American historian C. Dallett Hemphill puts it starkly: 'The whole purpose of manners was to silently accomplish the exclusion of the poor.'[23]

Moreover, when it comes to the special treatment of women and rules around 'ladies first', it can be argued that it was all a big charade. For Hemphill, all the codes of deference – the fact that it was up to women to initiate greetings, the elaborate rules on bowing, removing hats and so on – were in fact designed to uphold the existing patriarchy. In reality, women's social privileges were a way of compensating them for their ongoing political and economic marginalisation. And, of course, the underlying sentiment of 'ladies first' was not that women were mentally superior but that they were the 'weaker sex', to be protected by men and treated with extra delicacy. At a time when politicians were talking equality, the appearance of such chivalry enabled men to live up to their word, while privately maintaining the old view espoused by Chesterfield: that women were mentally incapable and best treated as sprightly children.

––––––

On reflecting on the history of etiquette – as the rise of inhibition, combined with an exercise in class control – it's tempting to reject everything the advisors have to say. Nowadays, followers of the principles of 'ladies first' and 'How do you do' risk being branded chauvinist snobs. This stems back to the countercultural movements of the 1960s and 1970s, which saw a rejection of social hierarchy and, with it, all the empty, suffocating rituals that the etiquette experts had preached. Instead, the rallying cry was for genuine equality and individual expression. Symbolising the break, people's behaviour and appearance changed. Traditional forms of address gave way to everyone being on first-name terms, while the morning suit was ditched in favour of jeans and a T-shirt. To use Randall Collins's phrase, it was the dawn of an 'era of casualness'.[24]

We can draw certain parallels between nineteenth-century Britain and America and what's going on in China and India today. Having experienced rapid economic growth for the past few decades, the two most populous nations are now seeing a rising middle class, along with deepening inequality. While the top end of society has developed a taste for luxury brands and is busy refining its social skills, the hundreds of millions of people at the bottom are still living in poverty. Viewed against the history described above, the boom in etiquette starts to look like an exercise in class consolidation – a wolf in sheep's (cashmere) clothing.

Yet, reflecting on my meeting with Hanson, I wondered whether all this was a bit harsh. For all his snobbishness, he seemed like someone who *did* have a wider social conscience. In fact, as part of his trip to India and China, he'd made a radio documentary in which he himself tried to understand why British etiquette has a growing appeal in such far-flung places. There is a moment when, sitting in a taxi in the choking traffic of Mumbai, he looks around and concedes that, for most people struggling with their daily lives, knowing how to fold a napkin is not a priority. And, for all her haughtiness and obscure rule-making, I'm guessing that the same could be said for Post. While appealing to high society, I don't think that either Hanson or Post could necessarily be accused of engineering class division. Whatever the reality, their central credo was and is that etiquette is for everyone. Putting aside some of the more arbitrary rules and dictates, there remains a basic message of respect and consideration for others – above all, this is what Hanson means when he says that etiquette and manners are indistinguishable.

And it could be argued that, at some level, etiquette is making a comeback. It's a movement characterised by quiet concern and private murmurings. It's people muttering about the decline in standards, complaining that we're losing our common courtesies. Or it's the feeling of unease and uncertainty about the basic rules in society, when to hold a door open or offer a seat – or how to greet someone. Without a popular uprising, it manifests itself in more subtle ways, in the way we tend to look back with rose-tinted glasses to a time when social life was more ordered: a recognition that for a free society to function there must be some constraints and boundaries. But, beyond that, it's perhaps an unconscious realisation that it's our social rules and codes that bind us. We depend on our rituals, conventions and routines to relate to each other. As the hippy communes discovered, informality and individuality were not necessarily the answers that would provide social happiness and harmony. When one set of rules and conventions is rejected, they usually get replaced by another. We might be living at a time when casualness rules, but that's the point: casualness has its own rules. Just as it would have been a social faux pas to turn up at a party not wearing a suit and tie in the 1930s, in some circles today it would be an equal blunder to turn up wearing one. In the same way, saying 'How do you do' would bring snorts of laughter.

In many ways, all this brings us back to the sociology of everyday life. Although writing at a time when the countercultural movements were in full swing, Erving Goffman didn't dismiss the rules of etiquette. In fact, he carefully studied them, using various guides as research material for his own investigation into the origins of social order. While he rejected their claim to any special authority on human behaviour, Goffman recognised that the etiquette advisors were essentially describing a set of social norms – capturing and codifying what we already do.[25] They were merely one set of conventions on which social order depends. As Adam Kendon found in his study of greetings, the way people said hello largely matched the advice of the latest Emily Post guide, though this did not necessarily mean that any of his partygoers had read it.[26]

As Goffman observed, all these unspoken rules and rituals reflect the particular circumstances and social context we find ourselves in, or what he called 'situational proprieties'. By and large, we adapt our behaviour according to how 'tight' or 'loose' a situation is.[27] So when we meet friends or family in a bar or at home, we tend towards more relaxed or looser exchanges. Conversely, if we are attending a funeral or job interview, we tighten our behaviour accordingly. Paradoxically, researchers have found that some of the tightest codes occur in places designed for relaxation. In his study of outdoor hot-tub usage (you did read that correctly), the Icelandic sociologist Örn Jónsson found that pool-goers tended to interact with minimal touching and conversation. As a rule, greetings were limited to a nod, even among those who had been frequenting the same tub for years.[28]

Although the rules we operate within might be largely unwritten, this does not mean that they are any less fixed or important. Working at around the same time as Goffman, another renowned sociologist of everyday life, Harold Garfinkel, showed just how much our interactions rely on unspoken assumptions and conventions. Recruiting his students to do the dirty work, Garfinkel instructed them to disrupt the normal patterns of behaviour. If someone said 'Have a nice day', the students were told to respond with 'Nice? In what sense exactly?' or 'What part of the day do you mean?'[29] Many of the experiments followed a similar pattern, with the subjects becoming upset and outraged after short periods. It's at this micro level, among the trivialities of everyday life that we don't usually think about, that social order begins. And it's when we break these rules and fail to match social expectations that things get awkward. If people don't know or violate these standards, they can become figures of fun or at worst excluded and isolated.

Beyond the 'social proprieties' of a particular context, our greetings tend to follow a number of rules without which they would become unmanageable. First, there's what Goffman calls the 'attenuation rule', whereby we modify our greetings according to how often we see someone.[30] After a long time apart, we tend to greet people with extended displays of the sort I observed at Heathrow. But to keep our daily lives manageable, we scale back our greetings as we see people more often. With our colleagues, friends and family, whom we interact with each day, once we've greeted each other in the morning, some form of the 'distant display', whether a passing 'Hi' or nod, is usually enough.

Second, and in a similar vein, there are the rules that determine how we respond to the question 'How are you?' As Harvey Sacks observed, if we admit to feeling terrible, the rules of conversation dictate that the other person should enquire what the matter is. The problem is that many of our encounters are with people whom we would not ordinarily share this kind of information with or expect to care. And, even when meeting someone we are close to, we don't always respond with a truthful answer, since our days would grind to a halt.[31] As Sacks says, when it comes to greetings, 'Everyone has to lie.'[32]

Last, there are the rules of precedence or who goes first, which, as Sacks points out, is a 'fixed problem'. It's like meeting someone on a narrow path: to keep moving, someone has to stand aside. Typically, the solution is to follow some social hierarchy, with the person of higher status usually taking the lead. In Victorian Britain, whatever the reality, it was always ladies first. In pre-Civil War America, reflecting the most extreme hierarchy, slaves were not to greet white people unless they had been greeted first. Across the world, societies have devised their own rules, usually based on age, gender or class of some sort.[33]

Again, though, we need to apply the fuller view of rituals – as symbols and builders of solidarity – to understand some of the rules and conventions around our greetings. As well as managing our interactions, they transmit our emotions and signify our relationships. When meeting someone for the first time or who is our senior, we tend to take a more formal and restrained approach. But as Collins observes, our exchanges take place along 'interaction ritual chains', whereby we carry the memory of our previous encounters and seek to build on them, reaffirming our ties. It's why a second greeting with someone is usually warmer and more intimate, or at least this is what's expected. Ultimately, as with our rituals overall, we are seeking to channel our emotional energy, using physical contact, rhythm and synchronisation to enhance them. Overall, though, the touching remains symbolic and is carefully moderated and timed so as not to give the impression of sensory pleasure. As Collins points out, 'deep tongue kissing could not be used for greeting in-laws.'[34]

So where does all of this leave Hanson and the other etiquette advisors? In some ways, I can't help feeling that they're redundant. At best, they're merely relaying what we already do, collecting rules rather than setting them. As the microsociologists have shown, our conduct and, most importantly, our rituals, are determined by the particular situation and context. By definition, then, etiquette is an elusive art. By constantly looking back to the golden age, etiquette advisors are playing an endless game of catch-up. Moreover, by viewing behaviour as a mark of class, they risk ridicule, resentment even. And yet, if we dismiss their underlying message – that we expect each other to follow certain social codes – we reject what holds us together and keeps society functioning.

All of which (sort of) finally brings me back to the moment I was standing in line, hoping to meet the Queen. She was being chaperoned by one of the senior members of staff, who was puffed out with pride. It seemed that she was being stopped in front of preselected people who'd done something special. They would bow slightly or curtsey, only shaking hands when the Queen offered, just as we'd been told by the aficionados. As she came level I edged forward, forcing my smile as wide as possible, hoping it would draw her in. But she was guided past and the moment was gone. I watched as she carried on down the line and then noticed my friend Rupert sticking his hand out, breaking the number-one rule. Maybe he didn't know. He was sure to get in trouble. But to my amazement – and annoyance – the Queen stopped and offered her hand. 'Hi. It's very nice to meet you,' Rupert said, grinning like a Cheshire cat. The Queen smiled back and Rupert spent the rest of the day gloating about it.

The point of this story? Don't bother listening to the etiquette experts, at least not if you want to stand the best chance of meeting the Queen. Whatever their advice, even the Queen is flexible. In fact, as her former equerry has commented, she too has become more relaxed over the years. And, apparently, Prince Harry says that he prefers to greet people with a hug rather than a handshake, indicating the trend of our times. But I think that there was also something more fundamental going on when the Queen accepted Rupert's hand. Royalty or not, she showed that she was not above the basic rule that to have rebuffed Rupert's offer would have been to reject his status as a socially acceptable human being.

4

The Seductive Science of Body Language and the Quest for the Perfect Handshake

It's March 2015 and the body-language expert Judi James is giving a talk on personal impact in London. The audience, a mix of sales and business people, is rapt. After all, this is one of the leading authorities on the subject, with James coaching multinational companies and making a name on the UK's *Big Brother* as the chief behaviour analyst. Once a catwalk model, she still cuts a striking figure, with a long mane of white hair and standing tall on high heels. By way of introduction, she gets everyone to imagine a time when they've got something big coming up, maybe an important pitch, promotion interview or hot date. 'You're feeling nervous,' she says, 'and you don't know how to look or what to do.' Smiles of recognition light up. These are the defining moments in life, moments when we could all do with an extra hand. 'And then someone will say the JBY phrase, which stands for what?' she asks. Most people look puzzled or are perhaps too embarrassed to answer, but someone from the back eventually pipes up: 'Just be yourself.' *Of course*, I thought. It's what everyone advises and I end up saying; in the end, there's no point trying to be anything else. James takes a step back, arms open: 'It's the most pants advice that anybody could ever give!'

So this is the great claim of the body-language experts: that we can consciously change our behaviour to boost our impact.

To be fair, James wasn't saying that we should lie or try to be someone else, but that we can learn to create the right impression – we can be *more* than ourselves. 'Image,' she says in her best-selling book, is everything: 'Get it right and you will increase your status, pulling power and promotion potential.'[1] It's hard not to be seduced. In contrast to the etiquette advisors, obsessed with observing fixed rules of conduct, whether out of politeness or snobbery, body-language gurus promise to show us how we can make the best impact and get ahead in life. Crucially, it's not what you say but *how* you say it that matters.

These are some bold claims. But how credible are they? Just how important are image and first impressions? Can we really control what people think of us, beyond what we say? And what role do our greetings play? Can we manipulate these opening rituals to boost our impact?

The starting point for all this is not body language itself, but the psychology that underpins it. For social psychologists, a key concept is 'impression management' – the idea that we're all acutely sensitive to what other people think of us and do our best to be seen in a positive light. Or, as Shakespeare put it, 'All the world's a stage, and all the men and women merely players.' One of the first people to give this proper thought was one of our former guides, Erving Goffman. Prior to his investigations on public order, Goffman's first and most famous book was *The Presentation of Self in Everyday Life*, which, originally published in 1954, is regarded as one of the most influential works of the social sciences. For all its notoriety, it had humble origins, beginning with Goffman's PhD research on the tiny Shetland island of Unst, 200 miles north of mainland Scotland. As an intrepid graduate, then in his late twenties, Goffman set out to investigate the local social structures, spending a year and a half on the windswept island, hanging out at the community hall and attending as many social events as he could, even the odd wedding and funeral. To avoid arousing suspicion, the earnest-looking student told people that he'd come to do research on crofting, despite not knowing anything about it.[2]

Unable to speak the local dialect and maintaining a studied distance from his subjects, Goffman cut a solitary figure. But slowly, through his observations, he began to develop the ideas that would catapult him to global fame. Watching the islanders, he noticed how they lived their lives as if they were in a theatrical performance, continually presenting different versions of themselves according to the situation and who they were interacting with. He noted, for example, how the owners of the local hotel, despite coming from local crofting stock, adopted middle-class personas in front of guests, both in how they spoke and in how they decorated the place. In his book, Goffman developed the analogy, contending that we divide our lives between 'front stage' and 'back stage'. Just like an actor, when backstage we tend to be more relaxed or busy preparing for our public appearances – these are the moments when we're singing in the shower or slopping around in pyjamas. As soon as we're with others (on the front stage), we jump into our public character. Think of the waiter who, back in the kitchen, is slumped against the counter, complaining about the difficult customers, but then snaps into a charming smile as he re-emerges into the restaurant – or, for that matter, any meeting at work.[3]

In the 1990s, the hit BBC sitcom *Keeping Up Appearances* was built entirely around this premise. The lead character Hyacinth Bucket (pronounced *bouquet*, she insisted) was determined to be seen as upper class, despite her working-class roots. At times, keeping up the front became a military operation. Finally, though, after an exhausting day maintaining the facade, she'd shut the front door, slump into a chair and relax her vowels. While I, along with millions of others, loved the show, it wasn't exactly cutting edge. Yet it's become the BBC's biggest global export, more successful than the likes of *Top Gear* and *Doctor Who*. Part of the reason, I think, is that people all over the world can relate. Whether it's the way we dress, the things we say or the books we keep on our coffee table, we're all trying to control how others think of us. Consciously or not, to some extent we're all keeping up appearances.

Just as it might not take a social psychologist to tell us that we're playing up to the camera, we also know, or are at least told, that of all the impressions, it's the first that matter most. As the sayings go, 'First impressions are the most lasting,' and, 'You only get one chance to make a first impression.' Yet we also resist the idea. We like to believe that our important decisions are based on reasoned thought. 'Never judge a book by its cover,' our teachers tell us. But as much as we like to think that we're better, we know, if we're honest, that we are prone to making snap judgements. We do it all the time with people, perhaps most of all with those we are contemplating spending the rest of our lives with. 'Love at first sight' might be mostly reserved for Hollywood, but it remains a romantic ideal – that we'll know in an instant when we've met our partner in life. It's perhaps why Tinder is so popular: it does away with the pretence of anything else.

The Nobel Prize winner Daniel Kahneman has spent his career studying decision-making and human judgement. One of the most influential psychologists alive today, Kahneman has shown how, while we might be capable of considered and logical thinking in slow time, many of our decisions come from a much quicker system of thought, one that relies on emotion and intuition. Along with his collaborator Amos Tversky, Kahneman conducted dozens of experiments that showed how we're prone to silly biases, self-delusion and overconfidence.[4]

While Kahneman's work catalogues a huge range of faulty tendencies, for now, let's look at just one: anchoring – how we're often swayed by the first bit of information we receive, even when it bears no relation to the judgements we're being asked to make. Consider the results of an experiment in which judges in Germany were asked to pass a prison sentence on a fictional shoplifter. Before making their judgement, they were instructed to roll a pair of dice that, unknown to them, were loaded so that they always came out as a three or a nine. On average, the judges who rolled a nine said they would sentence the shoplifter to eight months, while those who rolled a three decided on just five months.[5] While unsettling, it's a trick I've used to good effect when raising sponsorship money – I always fill in the first amount. Similar sorts of experiments in which volunteers are unwittingly manipulated with dud bits of information have shown how we're prone to all sorts of errors, including making false deductions, searching for coherence, relying on small amounts of evidence and overlooking statistical realities. Reflecting on their implications, the economist Richard Thaler memorably concluded that 'rationality was fucked'.[6]

In showing how our thinking is riddled with these kinds of bugs, Kahneman has earned himself the reputation of being an insatiable pessimist. After all, his findings suggest that humans are often blind to reality. However, his moral is not that we should give up and go home, but that we need to be aware of such weaknesses in order to make good decisions. In short, beware of the gut and anyone who tells you that humans are entirely rational.

It's this kind of intuitive 'fast thinking' characterised by Kahneman that determines how we make our initial judgements of people. It was back in the early 1940s that the social psychologist Solomon Asch pioneered the systematic study of first impressions. Perhaps inspired by his early experiences as a Polish immigrant trying to fit into life in the United States, Asch became struck by how a mere glance and few words can be enough for us to create a story about highly complex beings. Working at Brooklyn College, he conducted a series of experiments in which participants were asked to form impressions of fictional people based solely on a list of personality traits. Even with limited and contrasting information, Asch's volunteers invariably formed unified views of his made-up characters, often creating mini biographies. Moreover, Asch found that just by changing the sequence of the words he got very different results. So, for example, participants formed a generally positive view when someone was described in the following terms: intelligent – industrious – impulsive – critical – stubborn – envious. But when the order was switched and 'envious' put first, they changed their whole view, describing the person negatively, as someone with serious personal difficulties.[7] This was an early demonstration of anchoring and the 'primacy effect', whereby we tend to put disproportionate weight on early information.

Building on Asch's work, many studies have since shown just how quickly we judge people. In the 1950s, it was found that impressions are formed in three minutes. Then, in the 1980s, Nalini Ambady, who was renowned for developing the notion of 'thin slicing', whereby we make big deductions from small bits of behaviour, considerably upped the ante. In one of her most well-known experiments at Stanford University, she asked students to watch ten-second silent videos of unfamiliar professors as they taught and then to rate them on a range of qualities such as honesty, competence and likeability. What was striking was not just that the students were able to make judgements based on a very limited experience of the professors, but that their responses closely mirrored the ratings by students who had spent a full semester being taught by them.[8]

More recently, a team of researchers from Princeton University's Social Perception Lab led by Alexander Todorov hit the headlines with their discovery that first impressions are formed not in minutes or seconds but in milliseconds – in other words, immediately. Todorov had volunteers look at photos of sixty-six actors, all wearing grey T-shirts and neutral expressions, and rate them against five traits: attractiveness, likeability, competence, trustworthiness and aggressiveness. Initially, they were shown the photos for a tenth of a second (less than the blink of an eye), then for slightly longer and finally for as long as they wanted. Strikingly, with the extra time, the students' ratings hardly changed – it merely confirmed their snap judgements. Moreover, the most consistent rating wasn't attractiveness, which you'd expect given that it actually relates to facial appearance, but trustworthiness.[9] As much as we don't like to admit it, we size people up at a glance.

Of course, the big question in all of this is whether these first impressions are accurate. Certainly, we all like to think that we're good judges. But, unsurprisingly, it seems that the truth is more mixed. Using a computer programme and people's ratings, Todorov generated a set of model faces matching different traits. So, for example, a more trustworthy face has higher inner eyebrows, pronounced cheekbones, a wider chin and shallower nose bridge, and vice versa (think Mikhail Gorbachev versus Vladimir Putin). And the more symmetrical and average a face, the better. It perhaps follows that we tend to find trustworthy faces more attractive. But while Todorov found that people's ratings are remarkably consistent, the relationship between facial appearance and personality is far from clear. Some have suggested possible links: between facial width and testosterone (a pilot study suggests that men with wider faces tend to have higher levels of the hormone, which is linked to increased aggression), or maybe it's our personality that moulds the way we look (eventually our faces reflect the expressions we use, so we end up wearing our personality), or perhaps it's self-fulfilling (consistently treat someone as untrustworthy or incompetent and they'll end up behaving that way).

Todorov remains circumspect, reminding us that consistency is not the same as accuracy. Some studies appear to show a link, while others suggest no relationship. Perhaps the striking thing, then, is that we routinely make these kind of judgements, even though they might not be accurate. It's not so much that we have superpowers of intuition but that we are prone to the kind of faulty thinking highlighted by Kahneman.[10] We're all amateur psychologists trying to explain the behaviour of others, but unlike the professionals most of us are too easily swayed by a whole range of irrational biases. One of the biggest culprits in all this is 'confirmation bias': we come to an initial view and then seek information that supports it, discarding anything that's contrary. It's what gets academics into trouble or, more seriously, can lead to detectives, juries and judges missing important evidence and making false convictions. What the Princeton experiments show is just how quickly this process begins. We make our mind up about people in a split second and then look for things that confirm our judgement.

Moreover, drawing on the findings of Asch, we know that we're inclined to form sweeping judgements from small bits of information, overgeneralising and then interpreting sub-sequent behaviour as the result of some underlying personality type. Conversely – and this reveals why we're such unreliable psychologists – when it comes to explaining our own behaviour, we usually look to external factors. It's what social psychologists Edward Jones and Richard Nisbett have dubbed the 'actor–observer' effect.[11] For example, if you see somebody being rude to a shop assistant, you will likely assume that they are arrogant or self-centred. Yet, when we find ourselves being similarly bad-mannered, it's because we're distracted or running late. As Kahneman concludes, the point of all this is not simply to despair at how we're all hopeless judges (and hopelessly judge-mental) but that we need to be aware of our tendency to make quick and irrational judgements so that we can start to make better ones.

Against all this, it's not surprising that greetings can cause such unease. While much is determined by how we read a face, Todorov has begun to show how we integrate other inform-ation such as expressions, posture and voice, forming a uni-fied impression. Moreover, our initial interactions – those first seconds when we're busily looking for information that confirms our first thoughts – are defined by our greetings. No wonder we worry about getting them right. Suddenly, the notion of 'greetings anxiety' starts to take on a new and very real meaning.

Taking anxiety as a whole, the most common form is social anxiety, which is caused by worrying about what others think

of us. Drawing on Goffman's thinking about impression management, the social psychologist Mark Leary developed what he
calls the 'self-presentation' theory. In short, social anxiety arises
when we want to make a good impression but doubt our ability
to do so.[12] In the extreme, our motivation to be seen positively
is sky-high, but for whatever reason we think that our chances
are zero. For actors, it's getting ready to walk on stage on the
opening night having realised they've forgotten their lines. For
the rest of us, it's those big moments (the important sales pitch,
job interview or first date) when the stakes are high but we've
lost all confidence that we'll be able to give a good account of
ourselves. To make matters worse, our bodies start to rebel. Our
heart rate increases, we begin to sweat and shake, and we lose
the power of memory and coherent speech. Our heightened
physiological arousal only makes us more self-conscious, feeding
our anxiety, and so the vicious circle goes on.

No doubt some anxiety is a good thing, motivating us to prepare for those things we're worrying about. But, in the extreme,
it can become debilitating, even self-fulfilling, leading people to
avoid social situations altogether. Such acute suffering isn't that
uncommon, with a recent survey finding that more than one in
ten Americans had experienced clinical levels of social anxiety
at some point in their lives.[13] And for most of us there are times
when we feel our nerves getting the better of us, when we find
ourselves under intense scrutiny, as if pushed onto Goffman's
'front stage' without knowing our lines.

So where do our greetings fit into this? Is 'greetings anxiety'
really worthy of the term? Certainly, among those suffering
from social anxiety disorder, greetings represent a very real
and common phobia. Looking on various support forums, one
sufferer confides: 'I can't stand going into work for this reason

alone... It's literally the most difficult part of my day.' Many others sympathise, admitting that they too had a mental block when it came to saying hello: '"Hi" is next to impossible.' But for many of us, even those who appear most confident, greetings can inspire a special dread. The cruelty of it all is that's it's often when we want to say hello most or make a good impression that our doubts get the better of us.

Of course, much of the issue with greetings is merely symptomatic of our wider anxiety about certain social situations, when the stakes are high and we know we're being judged – job interviews, presentations, first dates, approaching someone in a bar, and so on. It's by saying hello that we open ourselves to the scrutiny we worry about. Greetings, then, are merely the tip of the iceberg.

But when it comes to addressing social anxiety as a whole, these little rituals can play their own distinct role, for both good and bad. It's through our greetings that we break the ice and snap ourselves out of the spiralling doubt. We begin to make the unfamiliar familiar; strangers are no longer strangers. And, as with most things we dread, once we've got going, they're usually not as bad as we feared. In this sense, a greeting, particularly when approaching someone cold, can be the social equivalent of a strongman pulling a truck. To get the thing moving takes a Herculean effort, but once there's momentum it gets easier. Think of all of the times you've been anxious in a situation when you didn't know anyone, such as the first day in a new job or joining a club. It's only once we've said hello to a few people (a smile can be enough) that our nerves begin to settle and we start to feel better about ourselves. Even sufferers of social anxiety disorder reflect how the act of saying hello (acknowledging another person as a social entity) can bring a sense of relief and help them conquer social situations.

The fact that our greetings generally follow set routines can be a help too. It's in those first moments of interaction, when we're feeling most nervous, that we need a guiding hand. Our greeting rituals provide a template and little scripts to follow, relieving us from some of our self-consciousness and doubt. We can rely on the weather, a cheesy chat-up line or whatever script to get us going, all becoming actors of some sort. Of course, problems arise when these little acts come to an end. It's why we often end up using some sort of prop to connect us and keep us going, whether an activity, a fancy meal or a dog.

But these rituals can also be part of the problem. Having standard routines creates the expectation that there must be a right way to do things, certain rules and conventions that we should all know and follow, just like the etiquette advisors tell us. So we worry about getting them wrong and feel embarrassed when we sense that we've breached expectations. Moreover, our physical rituals often break normal barriers of physical intimacy, creating the potential for ambiguity and misunderstanding. All in all, we can easily end up following different scripts. Having watched multiple episodes of Channel 4's *First Dates*, I know that I'm not alone. It can be painful viewing, as handshakes get trapped between hugs and kisses are left hanging. There's an episode when East Ender Terry tries to play it cool when his date Levana turns up. First, he offers his hand and then leans in for the kiss, but he's still sitting on his stool, so ends up falling into her chest. 'You drunk?' she says. It goes downhill from there. These are what Leary calls 'self-presentation disasters'. Beyond the immediate torture, the worst thing about them, as Leary points out, is that they leave their mark, filling us with doubts for the next time.

How, then, can we break this cycle? How can we cure this aspect of greetings anxiety and remove the doubts? Perhaps the only failsafe option is avoidance: we can dodge all social situations and never face an awkward encounter again. Sadly, for many sufferers of social anxiety, this is a common path, leading to isolation and depression. But even for people who are more confident, uncertainty about the greeting itself can be enough to walk the other way or turn to our phone. A less drastic option is to abstain from the physical part, which generally causes most problems. It's an approach I've used in different forms over the years. If in a group, I might simply hang back until the formal greeting phase is finished or find something to hold on to (an extra glass or small child) that puts me out of action. Alternatively, another tactic I fall back on in emergency situations is to stop short and announce that I'm getting over a cold, which has the added advantage of showing how considerate I am. Unfortunately, though, it's not the best way to start a date.

More promisingly, we can try to resolve some of the uncertainty around these rituals. The common advice is: 'If in doubt, follow the other person's lead.' The problems come when everyone relies on this approach. A safer bet, then, is to become more confident in our own greetings. One way is to learn and practise our routines. As with anything in life, practice makes perfect, including mastering our social interactions. It might sound stupid and be awkward in itself, but practising greetings with friends and family can help. As for how to learn the routines, we can watch others – or consult an etiquette guide. Studies have shown that the mere provision of a script is sufficient to reduce anxiety and apprehension.[14] This is a rallying call to the likes of Hanson to give the rest of us simple and reliable advice. The problem here is that, when they're not coming up with arbitrary rules, etiquette advisors are often simply trying to keep up with the changing times. So, in the end, maybe we need to forget the idea that there is a right way and just be more confident about our own way. After all, it's amazing what you can get away with, if you're self-assured and open enough about it. Others will be most likely pleased to follow, relieved to be put out of their own doubts. But confidence alone won't overcome all of the 'situational proprieties' that we face. And, just as if we all follow, if we all lead on the basis of our own rules, our greetings will become unmanageable.

While we might not be able to solve the doubt, there is another approach to all this: treating the side effects of our anxiety, controlling the nerves and self-consciousness. To this end, the most common and readily available treatment is alcohol. Certainly, it explains why so many first dates start in a bar. Or we can get various prescription drugs such as diazepam or beta blockers, which can help lower our heart rate, though this may be a bit extreme if we're just trying to address some low-level greetings anxiety. Alternatively, and this could be the biggest help, we can take what we've learned from the social psychologists and console ourselves by acknowledging some of the bugs that infect our thinking and compound our anxieties – such as the belief that everyone else knows what's going on and we don't. Or there's what's known as 'protagonist disease' and the 'spotlight effect': the fact that we're all at the centre of our own world, and so tend to think that other people must be always thinking about us too, when the truth is that they're more focused on themselves and what everyone else thinks of them. As Eleanor Roosevelt put it: 'You wouldn't worry so much about what others think of you if you realised how seldom they do.'

So when it comes to greetings, the point is that we shouldn't be so put off by the thought of messing up. People care less than we think or at least are more concerned about what we think of them. It's a start anyway.

———

Beyond fighting the awkwardness, perhaps we can tweak these routines to make the best impression. This brings us back to body language: our expressions, eye contact, gestures and posture, as well as the tone, speed and rhythm of our voice. The body-language coach's favourite statistic, often coming at the front of their books, is that these 'non-verbals' account for 93 per cent of our communication. Of this, 55 per cent is said to come from our physical expressions and gestures, while 38 per cent comes from the way we speak. This leaves a paltry 7 per cent to the words themselves. So fixed and reliable is this formula that it's commonly known as the '7%–38%–55% Rule'. As the renowned American body-language expert Janine Driver concludes, 'We put all the significance on words but we shouldn't.'[15]

Yet all of this seems to defy common sense. When listening to the radio or speaking on the phone, do we really catch only 7 per cent of what people are saying? I've been sent on week-long courses ('Communicating with Impact' and 'Communicating with Impact II') where the instructor hammered on about the importance of body language, even recording us giving present-ations to demonstrate the fact. I understand the purpose – body language can help convey a message. But surely the starting point for all communication is having something worthwhile to say. Above all, confidence and conviction come through knowing your stuff and believing in it.

To prove others right, I often bring up one of my favourite history professors, Peter Hoffmann, an expert on the German resistance to Hitler, who is based at McGill University in Montreal, where I spent a year as an exchange student. On the surface, Professor Hoffmann did everything to make his lectures as unappealing as possible. For starters, he held them at 8.30 a.m. to put off anyone who wasn't serious. Every class was the same. He would stand behind a wooden lectern and read from his fading notes. His voice, still carrying a thick German accent, was slow and monotonous. The only time he became animated was when he did a frighteningly realistic impression of Hitler, fists clenched and spit flying everywhere. Professor Hoffmann took some getting used to, but I was riveted by what he had to say: his stories of Nazi Germany and those who had fought Hitler from within. He was renowned for inspiring loyalty and respect among his students. The point is that, in a world where we're told that 'image is everything', the likes of Professor Hoffmann stand tall for substance over style. It all makes the 93 per cent figure sound ridiculous, if not pure propaganda.

So is there any science behind this so-called '7%–38%–55% Rule'? In 1994, a trainer in linguistic programming, Buzz Johnson, also had doubts, having heard it quoted many times but never able to get an explanation of where it came from. Finally, he met a professional speaker who made his living giving sales seminars and used the statistic as a central part of his message, as many others have and still do. The trainer was able to give a rough pronunciation of the name of the researcher he thought had devised the rule. Following the trail, Johnson eventually discovered its source: Albert Mehrabian, a professor of psychology at the University of California, Los Angeles.[16]

In the late 1960s, Mehrabian and a small team had indeed set out to investigate the relative importance of verbal and non-verbal messages. Although Mehrabian's findings were based on scientific methods and published in respected journals, as Johnson (and I) discovered, it turns out that their interpretation has been extremely misleading. In contrast to the popular claims, Mehrabian's research had a more specific purpose and limited design. Rather than testing communication as a whole, his team investigated the relative importance of verbal and non-verbal cues in expressing our feelings and attitudes, especially when they are incongruent. So, in his first set of experiments, Mehrabian asked participants to listen to tape recordings of different sets of words, one of which indicated liking ('honey', 'dear' and 'thanks') and another disliking ('brute', 'don't' and 'terrible'). Each set was spoken three times with different tones to indicate a range of feelings. The results were clear: when judging how the speakers felt, the participants were influenced by their tone of voice more than the words.[17] In the next experiment, participants were also shown photos of the speakers with different facial expressions as they listened to the words. This time, the facial expressions were found to have the greatest impact.[18] And so, weaving the findings together, Mehrabian got his 7%–38%–55% formula and the body-language experts got their big claim.[19]

But Mehrabian's methods were not designed to make such a sweeping judgement (he only used three speakers and all were women). Still, his findings demonstrate the relative weight of verbal and non-verbal communication when they appear to be expressing different feelings. For example, if I say 'Hello, it's so nice to meet you' with a sarcastic tone and sneer, you are not going to pay much attention to what's coming out of my mouth.

However, most of the time our words and expressions work in tandem – our body language supports what we're saying. Of course, it can add great power to our words, as Professor Hoffmann demonstrated with his impression of Hitler. This was the finding of another experimental psychologist, Michael Argyle, who, while based at the University of Oxford, was working at the same time as Mehrabian. In his experiments, Argyle got participants to watch videos in which actors spoke whole paragraphs, varying their tone and body language from clip to clip. Like Mehrabian, he was testing how the observers judged different attitudes, though he was examining perceptions of inferiority and superiority. Argyle's results also suggested that non-verbal cues carry greater weight when it comes to communicating our attitudes. Perhaps more significant to our understanding of communication overall, he demonstrated that the real power of body language is when it supports our verbal message. When the two work together, he found that they can have 4.3 times as much impact as verbal clues alone.[20]

In the end, the story of how the '7%–38%–55% Rule' became so prominent is less about understanding our communication and more about the misuse of statistics. That said, for all of the limitations of Mehrabian's methods and the subsequent over-interpretation of his findings, they still carry an important insight: that when our verbal and non-verbal channels appear to be saying different things, it's the non-verbal signals that provide a more reliable guide to what we're thinking. In short, while we can choose our words carefully, our body language is harder to control. It was perhaps King James I who first highlighted all this back in 1605 when, according to Francis Bacon, he quipped: 'As the tongue speaketh to the ear, so the gesture speak to the eye.'

Today, it's this ability to read what people are really thinking that makes up a key component of the body-language profession.

In this regard, one of the most recognised authorities is former FBI agent Joe Navarro, also known as the 'human lie detector'. Having spent twenty-five years working in counter-intelligence and behavioural analysis, helping to catch spies and criminals, Navarro now makes a living sharing his knowledge more widely, writing books, advising businesses and even coaching poker players. His craft relies on the notion of 'incongruence', which Mehrabian's experiments did much to highlight. He's made a career of spotting where people's body language conflicts with what they're saying. It's what's known in the trade as the 'leak' or the 'tell' – the giveaway that seeps out between the words.

Although Navarro used his techniques to catch spies and criminals, he also showed how we can apply them to our everyday encounters, taking readers of his best-selling guide on a tour of the human body and its various giveaways. In searching for clues to what everyone is *really* saying, Navarro starts in an unlikely place: the feet. 'The feet,' he says, 'are the most honest part of your body.'[21] When we're feeling excited, impatient or nervous, our feet act as a conductor. It's where we get the term 'happy feet' and what made so many people jig on the spot as they waited for their friends and family at Heathrow. Conversely, if you want to know when someone's had enough of your company, also look down – if they want to leave, whatever polite words are coming out of their mouth, their feet will start to point away from you, if not to the nearest exit. The direction of people's feet can also be useful in gauging if someone really wants you to join their conversation. If their body turns, but their feet stay put, it's a sign that you're better off finding someone else to talk to.

As Navarro explains, you can also spot when someone's feeling genuinely relaxed and comfortable by the position of their feet: if they're standing cross-legged, which is an inherently unstable, posture, it usually means that they're feeling at ease; if their feet are planted apart, probably less so.

Moving up the body, the torso sends similar signals through what's called 'ventral fronting' and 'ventral denial'. When we want to see someone, we turn the whole of our body to face them. By contrast, if someone approaches us whom we dislike or are unsure about, we tend to turn to one side. As Navarro concludes, in a way that's sure to get people paranoid: 'When it comes to courtship, an increase in ventral denial is one of the best indicators that a relationship is in trouble.'[22]

Next come our arms and hands. Beyond the thousands of symbolic gestures we make (thumbs-up, the 'V' sign, a circle for OK, and so on), they are prone to more involuntary movements, which can be a guide to when someone's feeling nervous or threatened. For example, when we're filled with nerves, we often bite our nails, twiddle our hair or fidget with a pen, and find it hard to stop. These 'displacement actions' or 'pacifiers' serve to channel all of our extra stress and tension into physical movements. Similarly, if we're feeling vulnerable or challenged, we'll often engage in self-comforting behaviour, such as rubbing our neck or arms, even giving ourselves a hug. And if we're feeling really exposed, we'll use our limbs to create barriers between ourselves and what's bothering us, typically folding our arms or maybe hiding behind an object. The higher someone holds up their glass, says Judi James, the more they want you to go away. In his study of greetings, Adam Kendon identified a form of barrier signal when he spotted people performing what he called the 'body cross'. A favourite case study for the body-language coaches is Prince Charles. When he's out meeting a crowd or walking a red carpet, he often twiddles his cufflinks. I'd always thought that this was just what royalty did, but the experts interpret it as a combined pacifier–barrier signal. In the extreme, if someone starts wringing their hands, it can be a sign that they're feeling very uncomfortable and, if they start clapping (to get their blood flowing), it could be one of the last indicators you get that they're about to attack you – a tip I could have done with at school on a few occasions. With the hands conveying so much, it's no wonder that Navarro found that we tend to think that people are trying to hide something when they keep them tucked away.

Navarro's tour ends with what he calls the 'mind's canvas': the face. Generally, it's the part of the body we do our best to control, but it's also the most leaky. Like fixing a dam, it's hard to plug one hole without forcing the water out elsewhere. The psychologist Paul Ekman, regarded as the 'godfather' of body language, discovered that we can make over 10,000 distinct facial expressions, each of which provides a clue to what someone is thinking and feeling.[23] Starting with the mouth, there are certain unconscious giveaways that someone is unsure of themselves. Most obviously, if they've just let out a bit of information they'd wanted to hide, they will put their hands to their mouths, as if to put the words back in – think of a child who gets caught telling their younger sibling that Father Christmas doesn't exist. Or, more subtly, if we aren't sure of something or don't want the information to slip out, we often draw our lips together. As for the expression of our emotions, as discussed, the upward curve of our lips (a smile) is the surest way to tell if someone is happy to see us, or at least wants us to think they're happy. But the smile is a complex thing, with body-language expert Robert

Phipps suggesting that there are at least twelve distinct types.

Of course, our mouths are only part of the picture. As any portrait-painter knows, it's the eyes that are the window to our soul. It's the eyes that show feeling and intensity – get them wrong and you lose the person. Also, we're visual creatures, and it's the eyes that show where our attention is. Again, we will come back to this, but for now there are a few obvious signs that something's up. As a rule, when we're uncomfortable about what we're seeing or hearing we'll scrunch our eyes, blocking out what's offending us. Navarro tells the story of when he was helping out a company with their contract negotiations. He got the client to read it out loud, line by line, and noticed them squint when they came to a particular paragraph. It alerted him to the fact that something was up. It turned out that there was a problem and resolving it was worth $13 million.

But, as Navarro cautions, you have to be extremely careful when interpreting body language. When someone rubs their neck or folds their arms, it could be that they have an itch or are getting comfy, not that they're hiding something or can't stand the sight of you. The key is to look for clusters of behaviour and anything that breaks with normal patterns. Even after twenty-five years in the business, Navarro concedes that detecting deception is extremely difficult. Repeated studies have shown that, as a rule, most people, including judges, lawyers, police officers, husbands and wives – even FBI agents – have no better than a fifty–fifty chance of telling when someone is lying.[24]

But even if the claim that body language makes up 93 per cent of our communication is misleading, it can still potentially quadruple the impact of what we're saying, if used effectively. A famous illustration of this – one that has gone down in body-language folklore as the moment that the profession was really born – occurred on the evening of 26 September 1960, when John F. Kennedy and Richard Nixon went head to head in the first ever televised presidential debate. Today, it's hard to imagine a US election without these public spectacles and the chance to scrutinise the candidates' every move. But Kennedy and Nixon were the guinea pigs and the results were telling.

Going into the first debate, Nixon was confident that he would win. Already vice president, he was an experienced politician and debater, stretching back to his days on the school debating team. By contrast, Kennedy was still a relatively unknown and inexperienced senator from Massachusetts. With typical bravado, Nixon told his campaign planner that the only thing that worried him was clobbering Kennedy so hard that he'd hand him the sympathy vote.[25] But in the event, with over 70 million voters tuning in, many seeing the candidates in person for the first time, Kennedy was judged to have come out on top. As commentators would oft retell, footage of the exchange reveals where a key difference lay. From his side, Kennedy, though young, looked the part. He was smartly turned out in a tailored suit and had a healthy tan. He stood tall, feet planted on the ground, appearing poised and unflappable. Nixon, on the other hand, looked ill. He had a pasty complexion, made worse by a last-minute attempt to cover up his five o'clock shadow with some 'Lazy Shave' make-up. His jowls drooped and eyes were like tree hollows. He stood uneasily, shifting from side to side, sweating under the lights, eyes darting about. All in all, dressed in a baggy grey suit, which blended into the background on black-and-white television, he cut a ghostly figure.

To be fair, Nixon was not at his best. Having recently been in hospital with a knee infection and just completed two weeks of intense campaigning, he'd lost 10 pounds and was suffering.

But the contrast was striking and, although Nixon managed to spruce himself up for the subsequent debates, the damage had been done. Kennedy went on to win the election by a narrow margin, with analysts concluding that it was his image that had made the difference.[26] This may or may not have been true but, for many, the lesson from the first debate was clear. As the *New York Times* journalist Russell Baker wrote: 'That night, image replaced the printed word as the natural language of politics.'[27]

With Nixon's grey and shifting figure casting a long shadow, our leaders and politicians, ever more exposed to the media, have increasingly turned to the consultants and body-language coaches to get their image right and push them ahead in the polls. Of course, some have fared better than others. Bill Clinton and Tony Blair, for example, are often acknowledged as masters of body language, using their gestures and tone to great effect. One of Clinton's signature moves was to put his palm to his chest, as if pledging allegiance, to show that he was speaking from the heart, while Blair famously captured the emotion of the nation following the death of Princess Diana with his appearance before the press (though I always felt that that his cracking voice might have been a step too far). In contrast, Hillary Clinton and Gordon Brown struggled as they tried to step into, respectively, the shoes of their husband and predecessor. As she moved from First Lady to secretary of state to presidential candidate, experts observed how Clinton's body language went from being understated and subservient to displaying control and authority. However, this came at a cost: for many, she seemed cold and aloof or, when she tried to make an emotional connection, calculating and phoney. Whatever her experience and credentials, some believe that it was a factor that put many voters off and blocked her path to the White House. As for Brown, the change in image as he succeeded Blair seemed to be the kiss of death. Whereas he had built his following on being the antidote to Blair's spin, on becoming prime minister he suddenly became camera friendly, appearing in a new suit, with whitened teeth and grinning in a way that no one had seen before. It looked suspiciously like the stylists had been brought in and instructed the man of substance to smarten up and smile more. The problem was that it didn't look natural; rather than boosting his popularity, it invited ridicule and eventually became part of his undoing.

James stresses the point in her March 2015 talk, as she picks on various politicians. With the general election leadership debates coming up, she shows a picture of Ed Miliband as another example of how body-language training can backfire. As she explains, while Miliband wasn't endowed with natural gravitas, he had his own charisma and geeky charm that could have been harnessed. But, at the cost of £130,000, Obama's advisors were brought in to boost his image. Rather than looking presidential, he ended up seeming awkward and staged. As James points out, one of the big giveaways that someone has consciously changed their body language is when their gestures and expressions come after the words (Brown's smiles, for example). To seem genuine, they should come a split second before, showing that the words are expressing our feelings. The mistake is what the experts call 'overcongruence' – it's not that people are lying, just that they're trying too hard and risk coming across as insincere.

Despite the potential pitfalls, James maintains that working on our social signals can give us an instant and easy advantage. When it comes to greetings, these are the moments that define what people think of us, whether they trust and have confidence in us. Drawing on the latest findings of social psychologists, she says that we make up our minds immediately and then scan for information that confirms our initial view. But while we might be stuck with our basic facial characteristics, we can at least control what we say and do.

So how do we make the perfect entrance and use our greetings to the best effect? As James, along with the rest of her profession, stresses, the key is confidence: 'Once you dither during a greeting, you're dead in the water.' As we've seen, for a variety of reasons, knowing which greeting to go for does not always come easy. But for the sales reps and business people whom James addresses, the one that's key is the handshake. Across the world, it's become the most common way to open meetings and seal deals. On the basis of rough calculations, it's estimated that most of us will shake hands around 15,000 times in our lifetime.

For body-language experts, getting the handshake right is crucial for making a good first impression. According to Navarro, it can define our opinions and relationships.[28] It certainly supports the popular view that you can tell a lot about someone by their handshake, with business leaders claiming to have decided whether to hire someone just on the strength (literally) of their grip. It's the subject of lively debate but, as a rough guide, a weak handshake (the 'wet fish') is widely interpreted as a sign that someone is submissive, nervous, shy or even sly and dishonest. While, at the other end of the scale, a heavy handshake (the 'bone-cruncher') suggests that someone is overbearing and arrogant – or maybe just insecure. It might seem like a crude measure, but there's no doubt that many of us are susceptible. If I'm honest, so am I. When I receive a real crusher, I can't help but assume that the person drives a high-spec BMW and likes to take charge in group situations, whereas, if I get a particularly soft one, I'll wonder if everything's OK.

Yet, however pervasive and important, it seems that many of us aren't getting our handshakes right. Take, for example, Indian prime minister Narendra Modi, who's been described as having a 'deathclamp'. Prince William was left gritting his teeth as he posed during their meeting in April 2016. A follow-up photo showed why: Modi's handprint was clear to see.[29] Or there's Donald Trump, who, despite his aversion, has become renowned for his 'yankshake', pulling his partner in and shaking vigorously – for a full nineteen seconds with Japanese prime minister Shinzo Abe. A recent survey suggests that, in truth, around 70 per cent of people in the UK are not confident about their own handshake, while nearly a quarter of people polled in my own region, East Anglia, would like to get rid of them altogether.[30] By far the biggest turn-offs are sweaty palms and loose grips, followed by a range of other pet hates including gripping too hard, holding on for too long, not making eye contact and shaking with the left hand.

Taking matters into their own hands, the US car company Chevrolet commissioned psychologists at the University of Manchester to find the perfect handshake. To much media hype, Professor Geoff Beattie, who led the study, came up with a mathematical formula that broke the handshake down into twelve elements: eye contact (e); verbal greeting (ve); smile (d) – the d standing for 'Duchenne smile', that is, one that includes both eyes and mouth; completeness of grip (cg); dryness (dr); strength (s); position of hand (p); vigour (v); temperature of hand (t); texture of hand (te); control (c); and duration (du).

Here's the formula in full, where 'PH' is the perfect

handshake:

$$PH = \sqrt{(e^2 + ve^2)(d^2)} + (cg + dr)^2 + \pi\{(4<s>2)(4<p>2)\}^2$$
$$+ (vi + t + te)^2 + \{(4<c>2)(4<du>2)\}^2$$

As impressive as it looks, for those who are mathematically literate, the formula was less than perfect. Moreover, critics point out that while it might be a great equation for exploring the gathered data, it doesn't necessarily tell us anything about the phenomena that are being explained, which could be entirely subjective. In the end, for all its problems and apparent complexity, Beattie's formula merely pulls together what the body-language coaches have been telling us. Based on their advice and in plain English, the perfect handshake goes something like this:

- Before shaking hands, make sure your palm is cool and dry. Wiping your hands on your trousers as you go to greet someone is a big turn-off.
- As with any greeting, smile naturally and make eye contact (holding it for around three seconds).
- Standing around 16 inches opposite the other person, extend your right arm at waist height so that your hands meet halfway.
- Make sure your grip is full and keep it vertical so your thumb is on top. Going underneath indicates subservience, while tilting your hand over the top suggests that you're trying to dominate.
- As for strength, go by the Goldilocks rule: not too weak and not too strong, but just right. If you feel a mismatch, adjust your own grip to equalise.

- Shake two or three times for around three seconds, keeping the rhythm smooth and controlled, while maintaining eye contact and smiling.
- Finally, release and step back, letting your smile fall away gradually rather than abruptly; otherwise it will seem forced.

All being well, this should match your fellow greeter's handshake. And, if you manage to pull it off, you can be pretty sure that you'll have won their trust and confidence. Do anything different, such as going for an alternative style or adding some sort of amplifier, and you'll likely put them off for good. Any extra pats will just seem patronising – as James puts it, patting is what we do to dogs and children. One final tip: if you're being photographed, try to position yourself on the right, so that your palm is facing inwards; that way, you'll appear to have the upper hand. Watch the likes of Vladimir Putin and other world leaders and you'll see them jockeying for position so that they get on the right-hand side and seem in control.

In the end, then, much of this comes down to the old-fashioned fatherly advice: 'Look 'em in the eye and give a good firm handshake.' But while there is nothing surprising about any of this, is it really true? Some people might well judge others by their handshakes, but isn't this nonsense? Isn't the notion of a perfect handshake just a convenient myth to keep the body-language experts going? Can you really tell if someone is right for the job just by the strength of their grip?

Handily (sorry), it turns out that social psychologists have wondered the same thing. In 2000, a team from the University of Alabama, led by William Chaplin, set out to test whether there really was any relationship between handshakes and personality. They started with a simple hypothesis, which was basically that the body-language coaches are right: handshakes that are stronger, longer lasting, warmer, drier and more vigorous, with a firmer grip and more eye contact, will result in a more favourable impression. The personal traits they tested for were extroversion, conscientiousness and neuroticism. To this end, they devised a cunning little experiment that enabled them simultaneously to test people's handshakes and personalities without them knowing what was going on. Just over 100 undergraduate volunteers, half of them male, the other half female, were invited to take part. They were told that it was an experiment to see if the way personality tests were conducted affected their outcomes. Each participant was asked to complete four questionnaires in which they scored themselves on a number of traits, including those that Chaplin was testing for. Before each round, they were met by handshake coders who had been specially trained to give a neutral handshake and measure the characteristics of the other person's. While the participants filled in their self-assessments, the coders graded the handshakes and recorded their first impressions.

The results were striking and unsettling, if not entirely surprising. Essentially, Chaplin's hypothesis (and the advice of the body-language experts) was proved right. Overall, the team found that the elements of a firm handshake co-varied. That is, people who had a strong grip also gave a vigorous shake, good eye contact and so on. And, crucially, there was a strong correlation between the coders' assessment of the handshakes and their assessments of the volunteers' personalities, as well as the results of the self-assessments. In short, those people with a firm handshake were judged by the coders – and themselves – to be more extrovert, open to new experiences and less neurotic. Moreover, the results were most pronounced for the women, allowing for the fact that they generally have weaker handshakes. Chaplin was emphatic: 'In summary, we found that women who are more liberal, intellectual, and open to new experiences have a firmer handshake and make a more favourable impression than women who are less open and have a less firm handshake.'[31]

Building on Chaplin's methods and findings, another team, led by Greg Stewart from the University of Iowa, pushed things further a few years later, testing the effects of people's handshakes in an interview setting. Broadly, the results matched Chaplin's findings: the participants who had the highest-scoring handshakes were also considered to be most hireable by the interviewers.[32] Taken together, the two studies seem to confirm what the body-language experts have been saying all along: we really do *and can* judge a person by their handshake. While neither study explains the correlations, as Chaplin reflected, it may be overstating it to say that a handshake is a window to someone's soul but, given the potency of first impressions, it's a good idea to make the first handshake a firm one.[33]

One interpretation of all this is that our personalities really do determine our handshakes and the powers to decode them are inbuilt. Although we haven't yet picked apart the causes of this unexplained power, psychologists have shown how people who are socially anxious often struggle to hold eye contact, so here's one legitimate inference at least: people who look away when shaking hands will tend to be more neurotic and less open to new experiences. Even this might be a leap too far, though.

An alternative explanation is that all of this is just another example of the kind of faulty thinking outlined by Kahneman. Essentially, for whatever reason – perhaps because it's what we've been told by the body-language experts – we're accustomed to the norm of a firm handshake; anything that deviates from this upsets our understanding of the world. And don't forget that we are prone to making snap judgements about people, forming an overall impression from little bits of information and creating an artificial sense of coherence.

Taken together, all of this amounts to what's been called the 'halo effect', whereby we quickly build a sense of the whole and keep adding to it, confidently interpreting people's behaviour on the basis of our underlying judgement. It seems likely that there's another bug at work: 'intensity matching' – our tendency to take different dimensions and relate them to the same underlying scale. For example, in a pioneering study, participants had different crimes described to them while listening to music, and were asked to match the volume to the crime. As the crimes got worse, people invariably turned the volume up. At a guess, most of us would be able to do the same with shades of red or green and people's confidence or aggression. Could it be that something similar is going on with our handshakes – the strength of grip and our perception of dominance, for example? Perhaps the real point that comes out of Chaplin's and Stewart's experiments is not that the handshake is a reliable guide to personality, but that we're easily swayed by arbitrary factors and false deductions – and tend to make some big decisions based on them, especially when time is limited.

All of this reminds me of an experience that's surely an exception that disproves the rule: my handshake with Henry Kissinger. It was October 2004, and Kissinger had been invited to meet a class from Yale University, where I spent a year of my PhD. There was a lot of secrecy around the meeting, since Kissinger is not a man who comes without controversy. From my own research on Anglo-American relations during his time in office, I was aware of the contested views: on the one hand, he's regarded as a shameless warmonger who extended the conflict in Vietnam and sanctioned the secret bombing of Cambodia; on the other, he's viewed as a peacemaker who helped to open relations with China and doggedly pursued peace in the Middle East, winning the Nobel Peace Prize for his efforts. Without doubt, though, this was a man who understood the power of image. During his time as Nixon's national security advisor, his whole reasoning for using force and, often, increasing the threat of force was that doing so would uphold America's reputation and deter the Soviet Union.

As he entered the room, escorted by our professors, Kissinger made a point of walking around the table and greeting each of us individually – here was a man who also clearly understood the power of a greeting. At eighty-one, he was greyer and rounder than in the books I'd been reading, but undoubtedly this was one of the 'Great Men' of history. As he got round to me, I stood up, though I found myself wedged against the wall and so was closer than the recommended 16 inches. First, there was eye contact and a smile and then the hand. This man had shaken hands with the likes of Mao and Brezhnev, so I'd expected an iron grip. But to my amazement I found myself squeezing what can be best described, in the words of Emily Post, as 'a miniature boiled pudding'. Here was the personification of power politics, yet he had one of the weakest grips I've ever come across. Kissinger went on to describe his time in office, recounting some of the defining moments in twentieth-century history, but I was left wondering about his handshake. Was it that he was submissive and neurotic? Or maybe it was a sign that he really was as sly and scheming as some have suggested? Or perhaps it was because he was eighty-one.

Kissinger's handshake stayed with me long after I finished

my PhD and accused him of undermining Anglo-American relations. Then, as I myself travelled and worked as a diplomat in China and the Middle East, I found that most people had a handshake like Kissinger. After all his travels and negotiations, then, maybe Kissinger had developed a handshake like theirs. At first, I have to admit that I found all the soft handshakes off-putting, forming judgements that I tried to resist. Put into a global and multicultural setting, Chaplin's and Stewart's findings, along with the advice of the body-language experts, become deeply problematic. How can we use the handshake as a measure of personality without dismissing most people in the world?

5

When in Rome

I t was the evening of 15 January 2015 and US secretary of state John Kerry was in Paris to meet with the French President François Hollande. He had come to show his nation's sympathy in the aftermath of the *Charlie Hebdo* attack. Lit up by the flashing cameras of the world's media, Kerry walked towards Hollande, wearing a sombre expression and holding his arms out as if getting ready for a hug. It was something that Hollande might have expected, since the night before Kerry had announced that the purpose of his visit was to 'share a big hug for Paris'. But with the hug not being a common greeting in France, local journalists had struggled to understand what he meant – the nearest word, *câlin*, has a romantic meaning and is closer to a cuddle.

As the pair met outside the Élysée Palace, Hollande put his own arms out and took Kerry's hands. It was unclear whether he was merely mirroring Kerry's gesture or guarding against what was to come. For a moment, the two statesmen stood opposite each other, hands clasped, as if about to waltz. Sticking to his word, Kerry released and went for it. Suddenly, he seemed to tower over Hollande, who, for his part, looked tortured. But as he managed half a hug back, he pursed his lips, getting ready for a kiss, as if trying to reclaim the greeting, French-style. Pulling back, Kerry noticed Hollande's move and also tightened his lips. But neither was fully committed so ended up jabbing at the air. It was like a failed smooch at a school disco, so tentative that they could at least deny anything had happened. Rather than of two men demonstrating their nations' solidarity, the lasting image was of two men separated by their cultures. For Kerry, what was intended as a sensitive gesture became a point of ridicule, with commentators mocking him for his social incompetence and blundering diplomacy.

Meetings between world leaders are littered with these moments. The two George Bushes were particularly prone. During a trip to Australia in January 1992, Bush senior was caught on camera flashing a V for Victory sign at a waving crowd. The only problem was that, being in Australia, he'd got his hand the wrong way around and was essentially telling them where to go. It was all the more unfortunate since Bush was on his way to give a speech promoting American–Australian cultural understanding. For his son's part, his attempts to play the Texan cowboy didn't always go down well on the international stage. As well as throwing Gordon Brown off by going diagonal, there was his infamous 'Yo Blair' moment at the G8 summit in St Petersburg in 2006. Even more cringeworthy, though, was the moment when, at the same meeting, he approached Angela Merkel from behind and gave her a shoulder rub, leaving the German chancellor looking like she was about to fall off a cliff. Even Obama, who has generally cut a more worldly figure, has not been immune to such blunders. When meeting Burmese opposition leader Aung San Suu Kyi, he went in for a kiss and hug, but she pulled back, looking faintly disgusted.

While these incidents symbolise the clash of cultures, mostly they've just resulted in personal embarrassment and something for the rest of us to laugh at. But a mix-up over greetings can have more serious consequences. For instance, there was the time when European explorers first arrived on the shores of New Zealand in the 1700s and were greeted by the Maoris with their traditional custom of standing up and raising their weapons. Perhaps understandably, the explorers misinterpreted this as a sign that they were about to be attacked, so fired at them, getting relations off to a frosty start.[1] Or there was the famous falling out during Britain's first diplomatic mission to China in 1793. After a year-long voyage to get there, expectations were high, with hopes of promoting trade and setting up an embassy in Beijing, but there was also disagreement about how the mission leader, George Macartney, should greet the Chinese emperor. The etiquette of the imperial court dictated that anyone approaching Emperor Qianlong should kneel three times and make nine full kowtows, touching their head to the floor. Despite the insistence of Chinese officials, Macartney refused, saying that he would only go down on one knee, just as he would before his own sovereign, George III. In the end, Macartney stuck to his guns, and the mission was a failure, with the emperor refusing British requests and demanding that George III respect his superiority.[2]

Clearly, for most of us, our day-to-day greetings are unlikely to spark an international incident. But the big difference today, compared with the end of the 1700s, is the degree to which people travel and mix with different cultures. Back in Macartney's day, he would have been one of a small number of diplomats, explorers and traders who travelled abroad. The notion of tourism barely existed. By 1950, though, around 25 million people were visiting different countries each year. Today, that figure is well over a billion, with one in seven of us spending at least one night a year overseas.[3] A major factor in all this has been advances in travel. When my great-grandparents emigrated to Canada in 1911 (having originally planned to take the *Titanic*), it took them the best part of a week to cross the Atlantic. At the end of the century, when I headed out to Montreal for an exchange year, it took me just over seven hours to get there (though my luggage didn't turn up for another week). Now, business people frequently attend meetings on different continents on consecutive days. Even for the most well-travelled, as well as jet lag, there is culture lag, as people catch up with the local conventions. More generally, the forces of globalisation mean that many of us work and communicate with colleagues from around the world on a daily basis.

The biggest driver of multicultural living has been migration. According to the International Organization for Migration, the number of people living in a different country to the one they were born in has increased by more than 40 per cent in the last fifteen years, with 244 million people (around one in thirty of us) now classed as international migrants.[4] The spread is uneven, but many of our cities have become cultural melting pots. The most diverse is generally thought to be Toronto, with around half of its residents coming from overseas, representing some 200 ethnic groups. London isn't far behind. Walk down Peckham High Street in one of the city's most diverse neighbourhoods and you'll hear accents from West Africa, Jamaica, India, Afghanistan, Colombia and China. 'Dave's International Hair Salon' is sandwiched between a halal fast-food joint and a Vietnamese supermarket, with the 'Persian Café' opposite.

With unprecedented levels of cultural mixing, for many of us the potential for intercultural mix-up is a part of everyday life. The big question with greetings is whose rules and rituals should we go by? Even if, as Kendon suggests, the sequencing is roughly the same from culture to culture, each stage brings the potential for mismatch and confusion. In a study of cross-cultural greetings, using foreign-language students as their subjects, a team from New York University uncovered layers of misunderstanding. Unsurprisingly, it was the physical aspect that presented some of the main difficulties. For example, the Japanese participants confided that they often felt highly embarrassed when enthusiastically hugged or kissed by Americans.[5] Conversely, talking to some of my American friends who studied in the UK, they found the comparative lack of hugging and smiling off-putting. On the flip side, people from cultures that aren't used to the extra level of intimacy, as well as feeling embarrassed, can develop misplaced expectations about a relationship.

Similarly, the team from New York found that what we say can cause problems. For example, the Russian and Ukrainian students were taken aback by being constantly asked 'How are you?', unsure why people they hardly knew wanted to know about their health. Some Americans said they were left speechless when met with the common Asian greetings 'Have you eaten?' or 'Where are you going?', misinterpreting them as invitations.[6] Overall, though, the researchers found that some of the main difficulties came in more informal settings or during chance encounters when the standard routines are dropped or adapted. Students from Hispanic cultures, for instance, said that they struggled with the sort of speedy, on-the-go greetings that are often used by Americans, finding them disrespectful. The subtleties in tone and phrase can also catch people out. Adding emphasis and 'so' to 'It's nice to meet you' will seem friendly in some parts of the world but smack of insincerity in others. While the difference between starting an encounter with 'Who are you?' and 'I'm not sure we've met before' can easily get lost in translation.

I could keep going, but the point is clear: our greetings are viewed very differently across cultures. The bigger issue, though, is the judgements that we make as a result. When we encounter unusual greetings, we don't just notice the differences: we inter-pret them. We make assumptions about people's core traits on the basis of the rituals and patterns of behaviour they grew up with. And we don't just judge individuals, but entire cultures – and often harshly. However misguided, we tend to overinterpret and look for evidence that fits our first impressions. One person I spoke to, who himself worked in a language school, recalled how, on his first day, when he went to shake hands with a girl from Saudi Arabia, she refused on account of her religious beliefs. He took it badly, admitting that it coloured his view of Muslim women.

So how should we greet someone from outside our own culture? When I put the question to William Hanson, his response was telling: 'When in Rome, do as the Romans do,' he said, before adding, with a twinkle in his eye, 'or as I like to say, do as the British do.'

Joking aside, here were two very different approaches to handling cultural diversity. We might think of those who go abroad taking their food and music with them, seeking out the nearest version of what they get back home, versus people who savour every difference and who come back dressed in whatever local clobber. As it turns out, these two conflicting attitudes reflect the changing approaches of the people who might be regarded as the real experts on culture and whom I was looking to for some guidance: anthropologists. For anthropologists, explaining human diversity has been a central endeavour, with their answers having profound implications for how the rest of us should understand – and greet – each other.

In many respects, 'do as the British do' has an older heritage, reflecting early anthropological thinking such as that of Edward

Tylor, whom we met in Chapter 2 with his attempt to catalogue and categorise the different greetings of the world. Appointed as the inaugural professor of anthropology at Oxford in 1896, at a time when the British Empire was at its height, bringing different peoples into contact as never before, Tylor is widely regarded as the first person to define and study 'culture'. While we tend to think of this in narrow terms, maybe reflecting a particular taste in art or music (what we call 'high culture'), Tylor took a broad view, describing it as 'that complex whole which includes knowledge, belief, art, morals, custom, and any other capabilities and habits acquired by man as a member of society.'[7] For Tylor, it was culture that accounted for all of our diversity, greetings or otherwise.

At the time, this was a radical view, with much anthropological thinking dominated by theories of race and eugenics – the belief that our differing behaviour was determined by our biology. Looking at the evidence, Tylor rejected this, arguing that our racial differences were superficial, disguising our essential sameness. Building on his analysis of artefacts collected from all over the world, from hunting tools and weapons to sculptures and musical instruments, Tylor observed how humans had come up with the same basic solutions to the same basic needs, deducing that we all share an underlying 'psychic unity'. Moreover, drawing on the theory of evolution, which had recently been popularised by Charles Darwin's *On the Origin of Species*, Tylor argued that every society was marching along the same cultural path. Like our bodies, he theorised that culture evolved from the simple to the complex, though, unlike Darwin, he argued that there was a direction to all this, with every culture advancing towards the same end through set stages of development. First, there was the 'noble savage': man at his most primitive, living in a state of nature. This was a simpler existence, a time when humans lived a life of subsistence, driven by natural impulses and superstition and lacking in intelligence and morality. Next came barbarism, characterised by agriculture, when humans applied their mental powers to cultivate plants and animals. Finally, and representing the highest form of culture, came civilisation, defined by science and reason.

Looking across the world and at all our diversity, Tylor contended that different cultures were simply at different stages along the same continuum. At one end were the hunter-gatherer tribes (the savages), with their primitive existence and mysticism. Rather than being evidence of our biological division, they were seen as a window into our common past. Many of the habits and customs he saw in his own society, from the jewellery people wore to the gestures they used, Tylor maintained, were simply leftovers from this early stage. At the other end of the spectrum, the advanced cultures of Europe and North America represented the high point of civilisation.

Most significantly for us, Tylor applied his evolutionary theory to greetings. By his logic, the differing rituals reflected different stages of cultural development. As Tylor observed in his *History of Mankind*, in general, primitive tribes used physical and expressive gestures, what he called 'pantomime', epitomised by their intimate and elaborate greeting rituals such as nose-rubbing, breast-slapping and the like.[8] These were what he categorised as the 'lowest class of salutations' and, in part, reflected the common view, taken from Darwin, that human speech had evolved from gestural language. It followed that such forms of communication were inferior and were therefore discouraged, even among the deaf.[9] Similar to animal greetings, these more physical rituals were about sensory pleasure. In contrast, having tamed their natural impulses, the Victorians exercised restraint, which became the guiding principle of their etiquette manuals. As one late-Victorian writer proudly proclaimed: 'We English... use gesture-language less than almost any nation upon earth.'[10] For Tylor, a swift handshake and 'How do you do' were not so much a badge of class as the very apex of culture.

At the time, this evolutionary theory of culture fuelled the imperial spirit, propelling the export of Victorian institutions, values and customs across the globe. While the empire might have disintegrated, for the likes of Hanson, the Victorians still set the standards of proper behaviour – at least, this is partly what gets him invited around the world. It's a view reflected in the response one of my friends got when she announced that she was about to start a new job teaching French: 'Isn't that English, only louder?' In reality, though, while the British might have had particular form, we've not been alone in harbouring a sense of cultural superiority. Just think of Emperor Qianlong, who regarded all the countries of the world as mere tributaries of China, or the rallying cry of US presidential candidates that 'America is the greatest', or, for that matter, pretty much any culture on the planet. Take the Dinka people of South Sudan, for example. The name Dinka means 'people'. Their bitter rivals are the Nuer, which means 'original people'. In many ways, when it comes to culture, Tylor's theorising gets at a popular view of the world: that our way is the best way. The corollary to all this is that our cultures create a sense of 'us' and 'them'. In the extreme, it's what academics call an aversion to the 'other' or, as Bertrand Russell put it, we 'fear and hate whatever is unfamiliar'.

Having become bound up in theories that justified racism and colonialism, anthropology went through an identity crisis, with a new school emerging that rejected most of what their forefathers had to say. Not only did they dismiss the idea that race determined people's mental make-up, but they discredited Tylor's whole notion of cultural evolution. While culture was seen as key to understanding human behaviour and diversity, all cultures were equal, or at least needed to be understood on their own terms. In this way, a member of a remote tribe was just as cultured as a Victorian gentleman.

The theory of 'cultural relativism', as it's become known, owes much to the ideas and experiences of one man: Franz Boas, or 'Papa Boas', as he was called by his students. Born in Germany to a Jewish family in 1858, Boas originally trained as a physicist, but developed an interest in geography and different cultures, taking part in expeditions to northern Canada to investigate the impact of the environment on the Eskimo people. Facing harsh conditions and often relying on the resourcefulness and generosity of the local population, Boas came to reflect on his place in the world, recording the seeds of thought that would eventually transform the way that anthropologists viewed the world: 'I often ask myself what advantages our "good society" possesses over the "savages" and find, the more I see of their customs, that we have no right to look down upon them.'[11] From this simple reflection, Boas went on to challenge theories of race, but also Tylor's ideas and the sense of superiority that went with them. Based on his experiences of living with different cultures, rather than simply collecting artefacts, Boas concluded that our cultural diversity came down to environmental circumstances and history. Where there were similarities in customs and beliefs, he contended that they were the result of independent invention and the spread of ideas rather than any 'psychic unity'.[12]

And so, applying this school of thought, we can start to understand all our particular greetings. For example, through his study of the Central Eskimos, Boas encountered many rituals that appeared to be unique to the region. It's how we get face-slapping. Among the eastern tribes, if a stranger arrived in a settlement, the locals would line up, with one man standing in front. The stranger would approach slowly, with his arms folded and head tilted to the right. The local man would then slap him on the right cheek as hard as possible, before letting the stranger have a go back. More rounds would follow until one of them submitted. All the while, the crowd behind would sing and play. However rough their own customs, the eastern tribes feared their western neighbours even more. Here, Boas had heard of visitors being forced into bouts of boxing, wrestling and something called 'knife-testing'. Perhaps most extreme, though, was the game of 'hook and crook', which was also common in Greenland. On arrival, strangers had to strip to the waist and sit on a large animal skin before being challenged by a local. The two men would clasp hands and try to straighten each other's arm. As Boas commented, the ritual was 'somewhat dangerous' since the victor had the right to kill his adversary. According to the locals, the purpose of these greetings was to find out who was the better man – quite different from anything Tylor had found.[13]

With Boas going on to set up the Department of Anthropology at Columbia University in New York, his thinking and methods had a profound impact on the discipline. Henceforth, anthropologists would become adventurers, immersing themselves in different cultures and studying their every detail. Culture was seen as its own sphere of analysis, entirely separate from biology or psychology. As one of his students, Alfred Kroeber, put it, culture is 'superorganic': self-perpetuating and transcending the individual.[14] Shaped by the environment, humans could be anything, unbound by biology or any fixed cultural path.

Following Boas's lead, another of his protégés, Ruth Benedict, developed the theory that to understand a particular aspect of a culture you have to understand the culture in its entirety. As she wrote: 'The whole determines its parts, not only their relationship but their very nature.'[15] It was a perspective that inspired a different kind of comparative approach among a new generation of anthropologists. Rather than highlighting com- monalities, as Tylor had, they took aspects of behaviour, many of which were assumed to be instinctive, and showed how they varied from culture to culture. One of the leading figures was Ray Birdwhistell, who in the 1960s pioneered the study of body motion and non-verbal communication, what he dubbed 'kin- esics'. Based on his cross-cultural studies, Birdwhistell concluded that there was no bodily movement, facial expression or gesture that had the same meaning across cultures. While nodding the head signals agreement in most parts of the world, in Albania and Bulgaria it's the opposite. Similarly, making a circle with your thumb and index finger may signify 'OK' in many parts of the world, but in Brazil and some areas of Germany it means arsehole. Or there are all our subtler movements such as lower- ing the body. Depending on the situation or culture, this could be interpreted as a sign of guilt, humility, amusement or that someone's getting ready to fight.

As Birdwhistell found, even the smile is used and inter- preted differently. While generally viewed as an expression of happiness, in some societies the upward curvature of the lips is a sign of embarrassment or malice. In the United States alone, Birdwhistell found that smiling varied from region to region, with people from the southern states smiling more than people from the north-east.[16] Subsequent studies have corroborated his findings, showing how since the 1970s Americans have become more smiley in photos and public settings. It's not that they're necessarily any happier, just that there's been a growing recognition of the social value of a smile, and as a result people are faking it more. It's even suggested that Jimmy Carter exploited this trend during his bid to become president in 1976, which saw him grinning from ear to ear. In turn, it's said that his wide smile may have prompted the broader trend towards more frequent and intense smiling.[17]

But all this smiling doesn't go down well everywhere. Recent research by the Polish psychologist Kuba Krys put to the test a well-known Russian proverb: 'smiling with no reason is a sign of stupidity.' Recruiting participants from around the world, Krys got them to look at photos of different people, both smiling and not smiling, and to rate them on intelligence and honesty. As he predicted, people's reactions varied from country to country. In eighteen countries, smiling actually led to higher scores on intelligence, though in six, including Japan, France and, unsurprisingly, Russia, individuals were perceived as significantly less intelligent when smiling. As for assessing honesty, most cultures associated smiling with trustworthiness, though there were some significant outliers, including Argentina, Zimbabwe and Iran, where the smiling faces were seen as less honest. When you put all this into the context of multicultural living and our use of mugshots, whether on job applications or dating websites, it starts to have significant implications.[18]

In a similar vein, a collaborator of Birdwhistell's, the anthropologist Edward Hall, investigated how different cultures experience personal space – what he called 'proxemics'. Drawing on an old idea, Hall contended that we are all surrounded by 'invisible bubbles', which dictate how we react to the presence and proximity of others. According to Hall, there are four bubbles.

Starting with the outermost, at 12 to 25 feet, there is our 'public' field. This is the distance at which we can communicate but can still do a runner if we need to. Second, at 4 to 12 feet, there is our 'social' bubble, in which we can have a conversation. Third, between 2.5 and 4 feet, there's our 'personal' space, which brings us within physical range, close enough to shake hands. Finally, within 18 inches, there is our 'intimate' bubble, which includes full body contact.[19] We might not be able to see these interpersonal bubbles, but we still sense them, particularly as they're crossed. Crucially, and as he'd expected, Hall found that the boundaries vary across cultures. What's public in one part of the world might be considered intimate in another. In particular, Hall highlighted how Americans and Arabs have very different spatial zones. Whereas Arabs tend to get up close in conversation, Americans would likely see this as an invasion of their personal space, and feel uncomfortable or threatened. This mismatch in requirements led some marines in Iraq to misinterpret the behaviour of locals, sometimes ending in a punch-up.[20]

Studies have shown how different cultures also use and respond to touch differently. Broadly, experts on non-verbal communication divide the world into high- and low-contact cultures, with the former found mainly in Latin America, southern Europe and Arab countries, and the latter mostly coming from northern Europe, North America, Australia and Asia. The psychologist Sidney Jourard put this to the test when he visited various cafes around the world and counted the number of times couples touched each other during the course of an hour. In general, his results confirmed the broad categorisations. He found that by far the most tactile were people from San Juan in Puerto Rico, who, on average, touched each other a full 180 times. Next, and bucking the northern European stereotype, though perhaps this was to be expected given its reputation as a city of romantics, there was Paris, with 110 touches. A long way behind came Gainesville in Florida, with just two touches. Unsurprisingly, bringing up the rear were Londoners, who didn't even get off the mark.[21] All this was back in 1966, so things have changed, but the overall point remains: people from different cultures are accustomed to different levels of touch.

Similarly, in some societies, eye contact is a sign of mutual attention and respect, while in others it can be considered rude or hostile. In general, studies show that high-contact cultures are most comfortable with a direct gaze, while people from low-contact cultures, especially in East Asia, find it disconcerting, even disrespectful. The Navajo Indians think that looking someone straight in the eye means that you don't believe what they're saying, while in Ghana some children are taught not to look adults in the eye on first meeting.[22]

The upshot of all this is that different cultures inhabit different sensory worlds. While getting up close, patting someone's arm and looking into their eyes with a big smile might be considered warm and friendly in one part of the world, it could be deemed inappropriate, threatening or predatory in another. As we become increasingly mixed up, it seems that we're doomed to awkwardness and misunderstanding.

———

If we apply the Boas school of anthropology to greetings, perhaps the only guiding principle and way to avoid embarrassment is to learn and respect local customs – in other words, 'do as the Romans do'. After all, the phrase and principle go back over

1,600 years to St Augustine of Hippo. Arriving in Milan to take up a professorship, he was puzzled to find that the church didn't fast on Saturdays like in Rome, so he consulted the local bishop, St Ambrose. Being older and wiser, Ambrose gave him the sage advice that would later morph into the famous saying: 'When I am at Rome, I fast on Saturday; when I am at Milan I do not. Follow the custom of the Church where you are.' But, if we go further and take on board Benedict's ideas, maybe we'll only get our greetings right by understanding each culture as a whole.

It's an approach I wanted to test, to see whether the way we greet really does reflect our entire cultural make-up. It seemed sensible to begin with my own culture, given that it's the one I know best and has caused me most problems. But immediately I stumbled into the issue of defining my culture and what level to choose. In today's world, our sense of belonging can be a complex business, with our cultures being increasingly connected and overlapping. Moreover, in my case, there's the whole British/English thing. Earlier I described myself as British, but strictly speaking this is to be part of a political union made up of four separate nations that, while having much in common, have their own distinct identities and history. By definition and design, Britain is *multi*cultural. So when it comes to describing my own culture, 'English' might be a more meaningful term, since it describes a nation that, while an invention itself, has existed in some shape or form for over 1,500 years. But there are still problems with this, not least the fact that the English themselves aren't always comfortable with the term. In a recent edition of the BBC's *Question Time*, the weekly programme that gives members of the public the chance to put questions to their politicians, a woman from Felixstowe talked about protecting 'English values and culture'. You could hear the rest of the audience take a breath. '*British,*' someone muttered. 'OK, British, then,' she said. For some, the term 'English' has become synonymous with nationalism and nostalgia.

But, to stick with it, it's hard to pin down what really makes English culture. If anything, paradox seems to be the hallmark. For some, it's about country walks, village fetes and unchanging traditions – 'England's green and pleasant land' – while for others it's about industrial cities, punk rock and cutting-edge science. We're famed for our manners. We fight to let people through doors and say sorry for saying sorry too much. But we're also renowned for having the worst hooligans. The caricature of the perfectly mannered gentleman stands side by side with that of the bare-chested bald bloke, wrapped in a St George's Cross flag, loaded with lager and battering some poor local with a plastic chair. We're often regarded as stuffy and conservative, headed by the longest-serving monarch, yet we gave the world the Sex Pistols and stormed the art world with an unmade bed and half a cow. We're known for being intensely private but, in reality, can like nothing more than a good gossip. We pride ourselves on fair play, yet our society remains riven by class division, deepened by inherited privilege and the most revered private-school system in the world. We celebrate being an open society with a global history and outlook, but we're also an island nation with a strong and vocal core of Little Englanders who'd rather keep ourselves to ourselves. I could keep going, but the point is clear: in nearly every respect, we are riddled with contradictions; we're essentially a bipolar nation.

Of course, much of this is crude stereotyping. In reality, most English people don't fall into such extremes but live their lives in shades of grey. We are, after all, also known for our moderation. The sense of paradox also reflects the fact that, like all nations, we're a living culture, endlessly contested and reinvented. When we look at greetings, the contradictions keep coming. In some circles, the double kiss is the new norm, while in others it's avoided on principle. You could get a handshake that makes you wince or one that will hardly register. Prince Harry might prefer the hug, but some of us baulk at the idea. 'How do you do' is still considered proper by some, while laughed at by others who are more likely to greet their friends with some form of abuse. The usual response to 'How are you?' hovers somewhere between neutral and negative, but you'll also hear 'Fantastic!' and even 'Bloody brilliant' (though there'll likely be a sarcastic tone). Of course, much of this comes down to our particular mood, the situation and the people we're meeting; but, like our differences overall, a lot is determined by factors that cut across English culture, such as gender, class, region and age. In short, a retired steelworker from Sheffield is unlikely to greet you with a double kiss and 'mwah mwah' (though he might call you 'love').

So is it useful, or even possible, to talk of an English approach to greetings? Are there any traits that apply as much in a rural village as in inner-city London, or as much in Cornwall as in Newcastle? To get some answers, I arranged to meet the anthropologist Kate Fox, who co-runs the Social Issues Research Centre in Oxford, though she is most renowned for her best-selling book *Watching the English: The Hidden Rules of English Behaviour*. During nearly two decades of research, Fox became a secret observer, trying to catch and make sense of every aspect of our behaviour. She even jumped queues and deliberately bumped into people, putting Englishness to the test. In the end, she concluded that there really is a coherent and distinctly English way of doing things. It's what she calls a 'grammar of Englishness': the unspoken rules, codes and standards that we follow instinctively.

'How should we do this?' Fox joked, as she opened her door. Clearly, it had been on her mind too. For the record, we shook hands but, as we'll see, the joking was an important element.

'Can I get you something to drink?' she asked, before disappearing into her kitchen and returning with a tray of tea and biscuits. 'You know, this would have been rude if we'd been in parts of Lebanon,' she said. As she explained, if food and drink is offered too early it can be taken as a sign that the host wants to get rid of you. It reminded me that there are downsides to being hyper-aware of our cultural differences.

I'd imagined that spending so long studying your own culture could make you either unbearably knowing or painfully self-aware, so was surprised to find that Fox seemed so unaffected and easy to talk to. To be sure, she was unmistakably English, full of self-deprecating jokes and apologising for everything. I was already getting a sense of the answer, but was keen to know whether watching the English had helped to cure her of the social condition that she'd diagnosed. 'No,' she smiled, 'but it does allow you to laugh at yourself more.' If anything, then, it had just made her more English.

Overall, talking to Fox was both liberating and slightly traumatic. It was like some weird therapy session, probing the depths of our national character, but also my own. It's with her help, though, that we can start to identify a number of rules that characterise English greetings. As she explains in her book, we should take rules in the widest sense: as a set of standards, norms and codes that are unwritten and unbinding, but nevertheless widely understood.

First, when it comes to greetings, there are no rules (at least not in the strict sense). As much as the likes of Hanson are doing their best to promote fixed standards, we don't have any. As a result, we're often unsure whether to shake hands, hug or kiss (how many to give and whether actually to kiss), or even what to say, and we don't expect others to know either. Curiously – and here's the paradox again – for a nation that can still pull off the most intricate ceremonies, when it comes to basic human interactions we can get completely befuddled. While other cultures have standard rituals, we're constantly working them out as if it's our first time. Working as a diplomat in Libya and Sudan, I was often struck by how we struggled with the basics. We could put on a fancy meal or big event, but when it came to everyday hospitality, we were strangely lacking. While our hosts would always receive us with a generous welcome, seating us in deep leather sofas, offering various teas and local treats, and making time for personal chat, we'd often struggle to rustle up enough chairs and a cup of water, before pushing things along as efficiently as possible. For a nation famous for its ceremony, our day-to-day interactions are notably unceremonious.

And yet, as much as we may seem allergic to rules and customs, we can be sticklers for them. For example, there was the moment when Jeremy Corbyn, newly elected as Labour leader, stood in front of the Cenotaph on Remembrance Day. No sooner had he bowed than there was a national debate about whether he'd bent low enough. As Erving Goffman and Harold Garfinkel would have concluded, our rules might not be written down, but they're no less binding.

Second, awkwardness rules. Obviously, this is closely related to the above. Without any fixed standards, we're prone to confusion and embarrassment, swimming about in a sea of uncertainty. Inevitably, this means that our greetings are often inept and clumsy, full of sidesteps and misfires, particularly with people we don't know so well. Often, the major culprit is the follow-up greeting, when a simple handshake doesn't feel enough, but there's no clear graduation. As Fox explained, to be truly English, you need to seem sufficiently muddled and self-conscious. And to properly nail it, you should throw in an apology or two. In England, awkwardness is visible. In fact, to Fox's amazement, some etiquette advisors have applied her insights to their own guidance. To get a greeting really right, you need to demonstrate that you're not sure what you're doing.

Third, reserve rules. Despite all the loosening of the 1960s, the old cliché is still true: both physically and emotionally, we are, comparatively, a low-contact culture. As the international-etiquette expert Dean Foster concludes: 'The basic rule is to minimize physicality.'[23] When it comes to our day-to-day greetings, often there's no physical contact at all. And when meeting people for the first time, our greetings can still seem Victorian, as we keep them brief and unfussy. Handshakes are efficient, with few extras (unless you're a politician), while the idea of greeting a stranger with a double kiss or hug fills many of us with dread. Our verbal greetings also tend to be low-key, giving away little. Hanson might be extreme with his rejection of 'How are you?' and 'Pleased to meet you', but the rest of us aren't far behind, often opting for lesser versions such as 'How's it going?' (dropping the questioning tone) and 'Nice to meet you' (said without expression). Any response to 'How are you?' that goes much beyond 'Fine, thanks' or 'Not bad' will likely seem jarring or ironic.

As Fox observes, in first-time meetings, particularly in an informal setting, there's also the 'no-name rule', whereby it's considered a bit forward to ask what someone's called or to introduce yourself by name.[24] We rely on others to do all that stuff. I was recently caught out by this when I turned up to my local running club for the first time, took a deep breath and began introducing myself to a few people: 'Hi, I'm Andy,' I said. One woman visibly recoiled. Instead, as Fox explains, the favoured approach in these situations is to wait until you're about to leave and then, as casually as possible, say, 'Sorry didn't catch your name,' or simply find out from someone else. All of which can lead to the ridiculous situation of groups of people who've known each for years, but still don't know each other's names and are too embarrassed to ask – like at my running club. As for asking what someone actually does, it's best to leave it for a few months, if at all. It's why, as much as we hate them, name badges and ice-breakers can be such a good idea among the English.*

Connected to our reserve is our understatement, which is also legendary. For example, there was the moment in 1982 when British Airways captain Eric Moody came over the tannoy as his plane started to fall, having gone through a cloud of volcanic ash and lost power: 'Ladies and gentlemen, this is your captain speaking. We have a small problem. All four engines have stopped. We are doing our damnedest to get them going again. I trust you are not in too much distress.'[25] In the same way, we tend to downplay the situation when things are going well. Stand outside an exam room or at the start of a running race and you'll mostly hear people declaring how badly prepared

---

* While our reserve might be evident in our greetings, there is one notable exception: how we greet our pets. Above all, the English are a nation of animal lovers. Around half of us own a cat, dog, gerbil or some other creature, with 90 per cent of us saying we treat them as one of the family, while some admit that they love their pet more than their partner. When it comes to greeting our animals, all physical and emotional restraint goes out of the window. We let our dogs slobber all over us and cradle our cats like babies, all the while declaring how much we've missed and love them. No doubt this reflects a genuine love of animals and the unconditional companionship they provide, but it's also partly the flip side of how we treat each other – a release for all the emotion that's trapped inside.

they are, how little they've revised or trained, even if they think they're going to blitz it. It's the same with our greetings. Even if we've just won the lottery, we're generally still only ever 'fine' or 'not bad'.

Our understatement is reflected in the fact that we can seem modest and self-deprecating. We tend to downplay our achievements and will do our best to convert any praise into self-mockery or palm it off onto someone else – just as Fox did when I said how much I enjoyed her book: 'Oh, well, you know, it's just pop anthropology,' she said, shifting in her chair. But as she explains, all this downplaying and self-effacement doesn't mean that the English are less arrogant than anyone else. We're just playing by our rules: the important thing is to *appear* modest. The problem is for outsiders who aren't accustomed to all this doublespeak and take us at our word.

Fourth, weather talk rules. It was more than 250 years ago that the British writer Samuel Johnson observed that 'when two Englishmen meet, their first talk is of the weather'. And, as Fox says, it's become customary for commentators to observe how this is just as true today. A 2011 YouGov survey found that we mention the weather on average once every six hours – no doubt often marking our opening exchanges.[26] As discussed in Chapter 2, this isn't necessarily because we're particularly interested in the weather or have a lot of it (though many of us are and we do), but more a reflection of the role that the weather plays in getting us to interact. It's something we all have in common and can be easily observed at all times, but, more importantly, it's safely unintrusive and doesn't require any emotional disclosure. There's nothing to stop us moving on to something more personal if we want. We're just getting going – the equivalent of a helping push when you're learning to ride a bike. Given the nature of our climate, a lot of our weather talk involves moaning about it. But, as Fox warns, outsiders should be careful, since the weather is treated like one of the family: while it's acceptable to complain about your own parents or children, it's bad form to pass judgement about someone else's (to their face, at least).[27]

Finally, humour rules. If there's one aspect of our national character that transcends our differences and that we're collectively pleased about, it's our sense of humour. We tend to joke about everything, especially ourselves. It's a basic social reflex, which pervades and sustains our interactions whatever the situation, whether in the pub or at a funeral. A bit like the weather, it's not the jokes themselves that are important, but the social role they play. They help to ease the tension and manage our awkwardness. But while our humour bonds us, in the extreme, our reliance on it can lead to an emotional disconnect, whereby we never really know what anyone thinks or feels. Instead, we just joke about it. In a way, then, our constant joking also reflects our emotional reserve and social inhibitions.

When it comes to our greetings, though, our tendency to make a joke can be a saviour. If we're not trying to say something funny, then the very fact that we're so bad at them gives us something to laugh about. Our confusion and awkwardness are ripe for mockery and self-ridicule. If someone unexpectedly goes for a double kiss, they can expect a sarcastic 'Look at you, going all Continental' in return. Or if there's a hesitation, someone might say 'Ooh, going in for a second', to which there'll be mutual laughter. Most likely, though, there'll be a pre-emptive apology: 'Sorry, don't know what I'm doing', to which the response will be 'Me neither' followed by more laughter. If in doubt, our hugs and kisses are laced with irony. When men hug, they'll often go in with exaggerated slaps to the back along with a wrestling noise, while the double kiss can come with an intentionally fake-sounding 'Mwah, mwah'. All of which helps to defuse their emotional content. There is a fine line, but, overall, our jokiness can help to relieve the tension at difficult moments. And it's our humour that makes our awkwardness bearable, even useful. So in a funny (ha-ha) way, you could say that our Englishness makes us some of the best greeters of all.

Of course, there are always exceptions and some of the above can be overplayed. Walk into any pub or watch *Big Brother* and all this talk of English reserve and understatement will seem hollow (though the alcohol can be a factor – which is perhaps the point), while people from the North pride themselves on being open and direct, and so on. But even if nothing is fixed, there are tendencies, norms and expectations. And though many of them apply to other cultures (especially the rest of Britain), they don't all come together in quite the same, unmistakably English, way.

All in all, then, it's clear that English greetings prove Benedict's point – that to understand any aspect of behaviour you need to understand a culture as a whole. In other words, in those first moments of interaction, our Englishness is often on full display. In the end, all our confusion, awkwardness, reserve and over-reliance on the weather and humour reflect what Fox diagnoses as the 'English social dis-ease', which she uses as shorthand to capture all our social inhibitions and incompetence. To outsiders it can seem at best curious, quaint and charming, and at worst nonsensical, off-putting and rude.

But the big question is: where does all this come from? How can a people who once believed in their God-given destiny to tell the rest of the world how to behave be so bungling in their own behaviour? It's a puzzle that many commentators have grappled with. Some point to our national obsession: the weather. While in sunnier climes people spend more time outside socialising in the street or park, we're often rushing to get out of the rain. Or perhaps it's our geography – we're a small and crowded island nation, and so put a premium on personal space and privacy. Yet many other nations share these environmental constraints without suffering the same social hang-ups. Some suggest, then, that it comes down to social factors, most likely fallout from our class system. We can end up acutely conscious about our relative social standing, unsure if and how to interact.

For her part, Fox is sceptical of any simple explanations. At best, we can put the English dis-ease down to a unique *combination* of factors, but no one can really explain it. But having trained as a historian, I'm tempted to search the past for events and periods that have shaped our national psyche. After all, looking back, it's clear that it wasn't always like this. When Erasmus visited England in the fifteenth century, he recorded with delight that 'when you arrive anywhere, you are received with kisses on all sides'.[28] And, as Kendon observed when showing me engravings by William Hogarth depicting scenes of everyday life, people in eighteenth-century London were draped all over each other. Of course, we can't go through all of English history here, but I think we can point to a few defining aspects of our more recent past that have changed things. Arguably, the overriding story of the last couple of hundred years, and something that's done most to shape our national consciousness, has been the rise and fall of empire. At its height, the British Empire was the biggest ever, covering a fifth of the world. And with it came a sense of superiority and authority. As Kendon suggested, reflecting on his own Englishness and greeting – or lack of greeting – with me, 'People in authority are kinesically much less mobile' (employing Birdwhistell's terminology). It helps to explain why the Victorians were so stiff or why the Queen can seem uptight (and why Prince Harry is more relaxed). In some respects, then, when it comes to our interactions and reserve, we're still suffering from an imperial hangover.

As well as loosening us up, the decline of empire also brought about a prolonged period of national self-doubt, captured by US secretary of state Dean Acheson's famous jibe that we've 'lost an empire and not yet found a role'. That was back in 1962, but we're still trying to reconcile a leftover sense of global responsibility with our diminished status, searching for alternative circles of power and influence, unsure which way to turn. On the one hand, we're pulled across the Atlantic, connected by our shared history, language and geopolitical outlook – our 'special relationship' – with the United States. On the other, we're pulled across the Channel to our closest neighbours, even if we've recently decided to cut some of our political ties. All of which is played out in our cultural life. We might lap up American culture, helped by our common language and love of US TV, but, going back to the court of Louis XIV and beyond, Continental Europe has always represented the height of sophistication to us. Yet, at the same time, we are wary of getting too caught up in either, bemused by the United States and suspicious of Europe. As we've looked across the Atlantic and over the Channel for a new role in the world, we've slowly adopted the hug and even the diagonal clasp from the United States, and we've borrowed the double kiss and *ciao* from Europe. But, ultimately, we're still not entirely comfortable with any of it.

Yet as much as the English mindset comes down to a range of negative factors (gloomy weather, crowded island, class division, national decline), we're not simply passive or reluctant victims of a social condition. In fact, much of it we secretly cherish (having a good moan, for example). And, without meaning to imply that it's all by design, when it comes to our inept approach to greetings I think that, underneath it all, it reflects a certain national resolve, pride even. So our lack of standard routines has something to do with the fact that we don't like rules – or at least we don't like having them written down or being told what to do – and are suspicious of dogma. While we might at times envy other cultures, with their rich traditions, we're generally wary of any kind of cultural nationalism and expression of it. Most of our own are laced with irony. And when it comes to awkwardness, we might not be the only ones, but it's something we revel in and celebrate. It binds and defines us. If someone's too smooth and self-assured when saying hello, we don't entirely trust them – they can't be one of us. (Then again, that might just be a product of some of our class divisions.)

As for our reserve, no doubt there's a degree of emotional frustration, but many of us are determined about it, maintaining that it reflects a sense of proportion, even self-confidence. It's the same with understatement and modesty. Even though it can seem like a big act (which it mostly is), I like to think that it reflects a certain realism about the world, even a consideration for others. It was the American writer Gore Vidal who said: 'Every time a friend succeeds, something inside me dies.' It can be how we feel when we receive one of those Christmas letters packed full of people's holidays and all the school prizes their kids have won, or trawl Facebook (which is basically an extended version): a bit worse about our own lives. As the writer Julian Barnes warned, the danger is that we judge the inside of our own lives against the outside of others. And so it is with our greet-ings. If someone tells us how fantastic they are, we feel slightly crushed. Our understatement and propensity to moan – 'Not too bad', 'Can't complain', and so on – guard against this. We might seem a bit glum, but we're also stopping each other from feeling bad about ourselves. That's my excuse anyway. I guess the point is that we can all come up with our own theories, none of which will entirely explain us, but our greetings will always reflect whatever one we concoct.

––––––

So, if that's the English, what about the rest of the world? Of course, we can't go through every country or culture explaining how and why people greet as they do. If we take language as a proxy, there are some 6–7,000 different cultures in the world, and this doesn't include all the local dialects. But with the help of anthropologists and international-etiquette experts, we can highlight some of the differences, the biggest pitfalls and causes of misunderstanding.

Before we consider our various rituals, there is the question of how likely people are to greet each other. As a visitor, it can define how we feel about a place. You hear it all the time: 'The people were so friendly' – or not, as the case may be. Often, it comes down to something as small as a stranger saying hello. Being English, I'm conscious that we don't always measure up well. But if we were to produce a 'propensity-to-greet index', we would undoubtedly be beaten to the bottom by the Nordic countries. Aside from being beacons of equality, the Nordics are renowned for seeming remote and taciturn. Commentators cite many stories to illustrate the point: Danes who miss their bus stop just so they don't have to ask their seat-mate to move aside; Swedes who'll take the stairs to avoid having to talk to people in a lift; Icelanders who don't acknowledge old school friends when back home.[29] Every culture has its introverts, but anthropologists have speculated about what's special about the Nordics. Maybe it stems from a Calvinist Protestant tradition that shuns brashness, or perhaps it's the cold weather that deters people from hanging about, or it could have something to do with their ethnic homogeneity, which means that there's an implicit understanding between them and therefore less need for interaction.[30]

Somewhere towards the other extreme is the United States, generally regarded as being more familiar and forward, where visitors can expect a cheery 'Hi, how are ya?' To the uninitiated, the extra openness can take them off guard, though it can also be liberating and certainly made my rides across America on the Greyhound bus more interesting. In contrast with the Nordic countries, it perhaps follows that a nation of immigrants should be more open and enquiring when meeting strangers. As many travellers have reflected, though, some of the most familiar cultures are in Africa. During my trip across the continent, it was common for people to stop and say hello, often asking where I was from, if I needed help, whether I was married and how many children I had. Of course, some of the attention came from the fact that, as a white guy carrying a big rucksack, I was obviously an outsider, but in many parts it's normal to greet strangers.

Some of it comes down to a difference between rural and urban areas. There's the scene in *Crocodile Dundee* when Mick Dundee, fresh out of the Australian outback, walks down a street in New York saying 'G'day' to everyone, only to be looked at like he's a nutcase. It seems that we have a common understanding: in our cities, we can't say hello to everyone or our days would grind to a halt, so rather than discriminate we say hello to no one. Conversely, even in England, it's common to greet a stranger when out in the country. At the extreme are the nomadic Tuareg people of the Western Sahara who might not see anyone for days travelling across the desert. But as a team of anthropologists found, when they do spot someone approaching, they *must* meet and greet them, even if it means going miles out of their way. With a history of intertribal feuds, not to do so would be threatening – though the Tuareg may also just crave the social contact.[31]

As Fox observes, what we might interpret as the rudeness or, at best, shyness of some cultures can come down to a distinction between what sociologists Penelope Brown and Stephen Levinson call 'positive' and 'negative' politeness. The former (the American and African way) is about making people feel included, showing an active concern and a need for approval, while the latter (the English and Nordic way) reflects a consideration for people's privacy and personal space, not wishing to impose or intrude. I used to see it all the time in Sudan. If a Brit arrived late to a meeting, they'd slide in, giving a sheepish smile and sit at the back, whereas anyone from Sudan would walk around the table shaking hands and saying hello to everyone, as is the custom. So the point is: when it comes to saying hello and interacting with strangers, we should realise that different cultures are accustomed to different kinds of politeness.

Turning to our various greeting rituals, we can also relate some of them to broader patterns and tendencies. Taking the verbal component first, most cultures have a single word or phrase that announces and acknowledges our presence, which roughly means hello. So we get *hallo* in German, *jambo* in East Africa, *ni hao* in China, and so on. There are some variations, such as *marhaba* in Arabic, which comes from 'welcome', perhaps reflecting the nomadic traditions, and *sawubona* in Zulu, which means 'I see you'. Many languages also have formal and informal versions, so in Finnish there's *terve* for strangers, *hei* for those you see more often and *moi* for the people you're closest to, while in South Korea the formal version is *annyeonghaseyo*, with *annyeong* being equivalent to 'hi'.

Across cultures, there are also the various forms of good wishes. Most commonly, they relate to a particular time of day, but there are also various peace gestures, such as the standard Islamic greeting *As-salamu alaykum* ('Peace be upon you'). The Qur'an instructs followers to respond in kind or with an even better greeting, so you'll sometimes hear Muslims trying to outdo each other, though a simple *Salam* ('Peace') can also work. Similarly, Jews say *Shalom aleichem*, though generally only when seeing someone after a long time. And, walking down Peckham High Street, seeing how many greetings I could find, I got a 'Peace and love' from a Jamaican shop-owner. Other forms of good wishes include *Tungjatjeta* ('May your life be long') in Kosovo and, my favourite, *Gamarjoba* ('I let you win') in Georgia.

Some cultures get more elaborate. The Yoruba people of Nigeria, for example, have a special greeting for every occasion: early morning, late evening, standing up, sitting down, when seeing someone who's pregnant (*A so ka le anfaani* – 'I wish you a safe delivery') and even going to the hairdresser (*E ku e wa* – 'Greetings for beauty'). The Kerebe people, who live on a small island in Lake Victoria, north of Tanzania, can get quite lyrical, particularly when men greet their sisters-in-law, usually with something like *Sula manchwanta galunga omugobe* ('Greetings to you whose saliva seasons sour vegetables').[32]

While cultures often mix and match, tailoring their greetings to different people, there's a notable exception: France – which happens to be the most visited country in the world. As the Canadian journalists-cum-anthropologists Julie Barlow and Jean-Benoît Nadeau describe, when in France there really is only one proper way to start an interaction: *Bonjour*. Whether walking into a shop or meeting the president, it's always the same. The French even have a saying for it, describing anything straight-forward as *simple comme bonjour* ('simple as hello'). Anything else, whether a cheery 'Hi!' or 'How are you?', just doesn't cut it and will likely get a snooty response. As Barlow and Nardeau point out, not knowing this golden rule can be a major cause of misunderstanding for visitors, generating some of the worst stereotypes. The point to know is that the compulsory use of *Bonjour* reflects a key aspect of the French national character, extending back to the revolution: the recognition of all citizens as equals. So, as Barlow and Nadeau advise: 'If it feels like you're saying it too much, that's just enough.'[33]

Next there are all the well-being enquiries, with an open-ended 'How are you?' being most widespread. The Iraqis have a more poetic way of putting it: *Shlonak?* ('What is your colour?'), while in Ireland you might be asked 'What's the craic?' – a term that captures the Irish love of banter. The Tuareg, on the other hand, take a more fatalistic approach, asking *Mā ijān labāsan?* ('What bad has occurred?'). Other cultures get more specific. So in China and much of East Asia, food is a favourite topic, with 'Have you eaten?' or 'Have you had your rice?' being common openers. Elsewhere, sleep is the big concern: the Quechua people of South America ask *Waráshcanqui?* ('Did you come well into the dawn?'), while further north in Mexico the native Tarahumara invite all sorts of responses with *Piri vi mure?*

('What did you dream?'). And in Mongolia, reflecting the harsh climate, it's common for herders to ask 'Are you wintering well?' or 'Are your sheep grazing in peace?'

In a number of cultures, particularly in rural Africa and the Middle East, these standard enquiries can take an extended form. The longest I've encountered was in a remote corner of northern Mali, inhabited by the Dogon people, who are famous for their distinctive art and, as I discovered, lengthy greetings. As I walked alongside my guide, Mamadou, each time we came within earshot of someone the questions would begin. Mamadou would enquire first after their health, then their wife's health, then their children's, then their parents', then their wife's parents' and, finally, their animals'. With the other person simultaneously asking the same, the whole exchange could sound breathless. To Westerners, who might throw out an 'All right, mate?' or 'How's it going?', this sort of prolonged interrogation can seem excessive, but they tend to reflect the harsher living of rural and nomadic cultures, so cutting them short would seem rude.

As we saw in Chapter 2, by and large these exchanges take the form of phatic communication, inviting a standard response – along the lines of 'Well'. At school, our French teacher would always start with *Ça va?*, to which we were taught to chant back *Ça va bien, merci*. Similarly, if someone from China asks if we've eaten, we should simply say 'Yes, thanks – and you?' or, if we haven't, that we'll be eating soon. And, if asked by a Tuareg if anything bad has happened, we should respond with 'Nothing – only peace'. There are, though, some subtle, but telling, variations. So while the English might respond with 'Not too bad', 'Can't complain' and so forth, in the United States you'll more likely hear an enthusiastic 'Great!' or 'Fantastic!', which gets at a broader difference: English understatement versus American overstatement. For the American part, it partly comes down to what's considered to be a core streak in their national psyche: feel-goodism. While most of us are at it, no other nation invests so much in fighting negative feelings, whether through therapy or antidepressants. So when Americans say they feel great, they want to mean it. But perhaps the most emphatic reply is the one I was encouraged to use in Sudan: *Mia mia* (which means something like '100 per cent') – and this in a country where there is a lot to complain about.

To be sure, not all these exchanges fall into the phatic camp. After months living in Samoa, the anthropologist Alessandro Duranti found that the most common greeting, 'Where are you going?', was a genuine question seeking information.[34] And, from my own experience of riding between *gers* (yurts) in Mongolia, the traditional greeting 'Have you tied your dogs up?' is far from meaningless.

Once learned, these phrases can smooth our cross-cultural encounters. Often, though, they open the way to some extra small talk, which is less scripted and can leave us at sea. Unsurprisingly, small talk causes a particular dread among the Nordics. The Swedes even have a special name for it: *kallprata* ('dead talk'). Instead, they can be more comfortable with silence. It's not for nothing that the hashtag of the Nordic Bakery chain of cafes in central London is 'myquietmoment'. Similarly, in Asia, there is a special reverence for silence. Rather than suggesting a breakdown in relations, it's regarded as a proactive form of communication. Reflecting their strong intellectual traditions, the French also tend to look down on empty chat. The key is to have something interesting or, better still, controversial to say. The Germans, on the other hand, tend to have a good moan, usually about the pressures of work. As a whole, northern Europeans are renowned for being contained at first. So, as a rule, even asking someone's name or what they do can seem forward. Understandably, all this reserve (negative politeness?) can seem unnecessary and off-putting to outsiders. Americans get around the whole issue of not knowing someone's name by asking. And in a nation that celebrates the self-made, asking someone what they do can seem like a natural progression. As described, in rural parts of Africa, the Middle East and Asia, visitors can expect lots of personal questions about their family and work. In general, though, as Foster describes, certain topics tend to be a safe bet when first meeting: personal interests, sport, complimenting the local food or someone's family; while it's best to avoid internal politics and passing comment on the country next door.[35]

For all the confusion that our verbal greetings can bring, it's the physical gestures that can get people into the biggest muddle. Though widespread, even a wave can backfire. In Greece, extending an outstretched palm is known as the *moutza* and is deeply insulting, dating back to a time when convicts were paraded in public and had dirt and faeces wiped in their faces, while in Mexico and Panama the same gesture is used as a threat. Similarly, you should be careful with the thumbs-up. A symbol of friendship in much of the world, going back to the Roman signal for sparing a defeated gladiator, in Sardinia it means 'Sit on this' and, in parts of the Middle East and West Africa, is equivalent to giving the middle finger. In Niger, shaking a fist at eye level is used to say hello, but in much of the Gulf gesturing with a closed hand is considered vulgar. And then there's the traditional Tibetan greeting of tongue-poking, which goes back to the ninth century, when an unpopular king, Langdarma, who had a black tongue, was thought to have been reborn, so people would use the gesture to prove that they weren't his reincarnation. It's still used as a mark of respect today.

When it comes to the close display, as we've seen, different cultures are accustomed to different levels of touch. So, along with the English, northern Europeans – especially the Nordics – tend to take a less tactile approach, mechanically sticking to the rituals. In much of East Asia, there's generally no contact at all, unless adapting to other cultures. Conversely, in southern Europe, the Americas, the Middle East and Africa, greetings can get extremely intimate, even with strangers.

But, as is often the case, the devil is in the detail. Even the handshake, which has become so universal, varies widely. As we know, in the West a strong handshake is widely associated with confidence and trustworthiness, with North Americans having the tightest grips. Yet the whole notion of the 'perfect handshake' falls apart when we consider other cultures. Far from being viewed negatively, across the Middle East and North Africa a soft handshake is sign of consideration and respect, while most cultures in Asia have only adopted the custom in the last 100 years or so. Even in Europe and South America, the strength of grip can vary from country to country.

Length can also catch people out. We might be accustomed to a brisk pump or two in the West, but in many cultures, from West Africa to the Middle East, people will keep shaking for the duration of the opening exchange. And as an extra sign of friendship in parts of the Gulf, men remain hand in hand as they walk off together. Then there are all the extras, which the English might distrust, but can be taken as a sign of sincerity elsewhere. In Spain, close friends and family often bring in the other hand to touch the elbow, while in Greece it might be used to slap the back or in Turkey to grasp the shaking hands and then brought back to hold the heart. All in all, the point is clear: when shaking hands across cultures, we should resist our judgements, even if scientific papers and etiquette guides appear to support them.

Beyond the standard handshake, there are all of the other versions, such as the diagonal clasp and fist bump, which are now common among friends in North America and creeping into Europe. Or there's what's sometimes called the 'African handshake', which is a three-part affair, starting normally, then going diagonal, before ending horizontal. And then the finger snap, also used across Africa, which involves using your thumb and index finger to grab the other person's middle finger, and then snapping them together. In Ethiopia they add a shoulder bump and in Sudan a shoulder pat. The list goes on.

Perhaps the most doubt and embarrassment, though, come with social kissing. Taking into account the rest of the world, the title of this book could have been *One Kiss or Two or Three or Four or Five or Six or Seven or Eight?* Yes – in parts of Afghanistan, it's eight. But we only have to take France as a microcosm of all our kissing dilemmas. From being toddlers, the French are taught what's affectionately – and less affectionately – known as *la bise.* The basic procedure is straightforward enough: starting on the right, pass from cheek to cheek, with a glancing touch, kissing the air (not the cheek); the accompanying noise should come from the lips (definitely not an ironic 'mwah mwah'); and, as Kerry discovered, there's no hug. But that's where any certainty ends. First, there is the question of whom to kiss. In general, *la bise* is performed between women, and men and women, though in some parts of the south it's also common between men. More tricky is knowing when a relationship warrants *la bise*; usually, it's a matter of being acquainted, but indirect acquaintances can also count, though kissing the boss is another matter. To prepare the way and avoid awkwardness, the French sometimes say *On se fait la bise?* ('Shall we kiss?'). But then the big question is how many. Whereas two might be standard in Paris, elsewhere it can be anything up to five: three or four in Provence, four in the Loire Valley and up to five in Corsica (though here they may also go for two, in which case they start on the left). The number can also vary according to age and class, with younger people and the lower classes tending to kiss more, though nothing is fixed.

And that's just France. In Spain it's two, usually between men and women, and women and women, though it can also be between men in the south. In Italy it's also two, though starting with the left. In Belgium, it's three. In Colombia, it's one. In Brazil, it's usually two, though a third might be added to a single person as good luck for finding a spouse. In Russia, it's two or three, followed by a hug. In the Muslim world, especially the Gulf, kissing is common between men, though never between the sexes. In the United States, cheek-kissing is spreading, but now there's a new custom: kissing on the lips. Then there are all the other variations involving different parts of the body: in Congo and Egypt, people kiss each other on the forehead, while in parts of Austria, Poland and Hungary, some men still greet women by gesturing a kiss to the hand. And so it goes on. Yet there are some cultures that don't use any form of social kiss – most of East Asia, for example. Or there are the Tsonga people of southern Africa, who didn't have any notion of kissing at all and were appalled when they first saw the habit: 'Look at these dirty people! They suck each other! They eat each other's saliva and dirt!'

Beyond the handshake and kissing, there are all the other physical rituals that endlessly vary from culture to culture. In a number of countries, the nose comes into play. There's the Maori *hongi*, for example, which involves pressing foreheads and noses together and breathing in – taking the *hā* or breath of life. A similar custom is used across the Polynesian islands, as well as by Arab men in parts of the Gulf, while people in Greenland and the South Pacific island of Tuvalu press their noses and lips against each other's cheeks and sniff. Or there are other forms of touch, such as pressing the temple three times with the index finger in Congo, putting the right knuckles against an elder's forehead on the Pacific island of Guam or touching the right foot of an elder in India.

In general, if we keep an open mind, the worst that can happen is that we get a bit confused and embarrassed. But where religion is involved, things can get more problematic. So in Buddhist countries, you should never pat someone on the head, since the head is considered sacred. In more conservative Islamic circles, all forms of physical contact between men and women outside the family are prohibited. Even the handshake is out, with the Prophet Muhammad being clear on the matter: 'It is better for you to be stabbed in the head with an iron needle than to touch the hand of a woman who is not permissible to you.' When I found myself suggesting to the imam at London's Central Mosque near Regent's Park that this sounded a bit harsh, he explained that it was about controlling temptation and protecting women. So, in general, men shouldn't offer and women should politely refuse. But the problem is that the handshake has become such a universal custom, particularly in professional settings, that Muslims and non-Muslims alike can easily be offended. Recently, in Sweden, a Muslim man was even fired for refusing to shake a female colleague's hand. Within Islam itself, the debate rumbles on. While many stick to a literal view, others take a more flexible approach, believing that God would understand. It's similar in other religions, with Orthodox Jews, for example, avoiding all physical contact, even eye contact between the sexes. It's no panacea, but as various etiquette advisors suggest, if you want to avoid causing offence or feeling snubbed when meeting someone from the opposite sex who's in religious dress, it's best to let them go first.

Many cultures avoid some of these difficulties by sticking to non-contact displays. Though no longer so widespread, the bow is still a common greeting in much of East Asia, extending back to the Confucian principles of deference and humility. Undoubtedly, the most prolific bowers are the Japanese, who regard it as somewhere between an art and a science, ensuring that children are taught the proper technique early on. To give his employees the edge, in the late 1970s the head of a department-store chain, Chiyuki Takaishi, even developed a bowing machine, with his most advanced version using electronic sensors to evaluate the three main angles of bow: 15 degrees (slight bow) for a friend or fellow employee; 30 degrees (medium bow) for a customer or to show extra gratitude; 45 degrees (deep bow) to show particular respect or make an apology. In general, it's proper form to bend from the waist and make sure that you go lower and longer than anyone of higher status, which can result in much bobbing up and down. In parts of West Africa, it's still common for men to kneel and scrape before their elders. And in much of South Asia, the traditional greeting is a bow accompanied with the prayer sign, while saying some version of *Namaste*, which means 'Bowing to you', reflecting the Hindu belief that God exists in everyone. But religion can still complicate matters.

In Islam, any form of bowing is strictly reserved for prayer – as a direct gesture to God. One Saudi cleric even issued a fatwa making bowing to anyone else an act of 'shirk' or false idolatry.

Beyond all the vocal and physical gestures, our greetings might involve a gift exchange, which can open up a whole other dimension of misunderstanding. We all like a present, but some cultures are more particular than others, with perhaps the biggest givers being the Japanese. When I visited Japan, I was instructed to fill my case with tea and biscuits to give to people. And when some Japanese students came to stay as part of the return visit, they brought all sorts of gifts, from paper cranes and fancy fans to photo books and calligraphy paintings. What I hadn't accounted for was the expectation of return, so had to stand by as they scoured my room, inspecting my most treasured artefacts (eventually, I handed over a couple of wooden statues my sister had brought me from Kenya). In Japan, present-giving is enshrined in the notion of *on*, which means 'what is owed', and reflects the deep sense of obligation that permeates society. The Japanese also take great care in the way their gifts are wrapped (usually in two layers and tied with ribbons), though they tend to open them once the giver has gone, thereby avoiding any display of disappointment and, crucially, saving face. In the Arab world, gift-giving is common in business situations, when specially made mementos are often presented. You should be careful what you admire, though, whether a watch or an ornament, since a more traditional host may feel obliged to give it to you and take offence if you refuse – though, again, reciprocation is the norm.

Across the world, you should also take care with what you give, since superstition is rife. So, in South America, giving scissors or a knife could be taken to mean that you're cutting off a relationship, while handkerchiefs connote grief. In China, clocks and straw sandals are associated with funerals, while anything wrapped in yellow paper with black writing is given only to the dead. In Mexico, Chile, Russia and Iran, anything yellow can be a sign of grieving or hatred. In Kenya and Tanzania, flowers are reserved for condolences. And, for religious reasons, it's best not to give a Muslim anything alcoholic or an Indian anything that's made of cowhide.

All in all, it's clear that the way we greet is determined by the cultures we come from and that our different conventions get at deeper tendencies. So, taking the anthropologists' lead, it's only by learning and respecting each other's customs that we can avoid embarrassment and giving offence. Yet, there's a danger here: if we get hung up on all our differences, trying to understand the ins and outs of every culture, we risk turning the world into one big social minefield and will never get anywhere. Moreover, as Foster warns, if we try too hard to fit in, we'll only end up seeming fake and insincere.[36] And, as Hanson pointed out, this is the biggest faux pas of all.

So perhaps the important point in all of this is not that we should get our greetings right, perfectly blending in, but that we understand our differences and try not to react negatively when we're surprised by them or if people get them wrong. After all, we are only human – each of us individuals with our own thoughts, experiencing the world in a different way, having good days and bad days, however much we're shaped by the culture we're from. In the end, our particular greetings are just the behaviours we've grown up with.

When it comes to exploring the different peoples of the world, one of my heroes is the TV adventurer and former Royal Marine Bruce Parry. While not formally trained as an anthropologist,

Parry takes participant observation to the extreme, living with different tribes and, as far as possible, becoming one of them. He's made 'first contact' with the Korowai tribe in Papua New Guinea, danced with the Babongo people of Gabon, run naked across cows in Ethiopia and hunted with the Penan of Borneo – all in the name of cultural understanding. Yet for all of his encounters with so many unique cultures, as he's sat around campfires at the end of the day talking to people about their hopes and dreams, his overriding impression is that, fundamentally, 'we are all the same.'[37] As I wandered the streets of London, conducting my own bit of participant observation, seeing how many different greetings I could find, meeting people from all corners of the world, the same message kept coming back to me. So could it be that our greeting rituals have more in common than we think?

6

The Genital Grab and the Evolutionary Origins of Greetings

I n my quest to find the most unusual greeting of all, to reveal humans at their oddest, I found myself typing 'genital grab' into Google. At the time, it was the most inappropriate way of saying hello I could think of. Unsurprisingly, an unsavoury list of stories came up. But as I clicked past the first few pages, I came across a thread mentioning the Walbiri tribe in central Australia, who were rumoured to say hello by shaking each other's penises. It was apparently the ultimate expression of goodwill, though most of the posts suggested that the whole thing was a wind-up. But I also found references to Sicilian monks doing something similar. It felt like a eureka moment. Surely this was it: the oddest greeting of all, behaviour that had no possible explanation in terms of my own culture. Yet, as we'll see, far from demonstrating some perverse impulse or uniquely weird rite, understanding this particular greeting will take us deep into our evolutionary past, revealing connections to the rest of the animal world and theories that unite our species as a whole. In fact, there's good reason to believe that we were all genital-grabbers of some sort at some point (and, in a way, still are).

As we know, it's widely suggested that many of our commonest greetings originated as means of demonstrating that we weren't armed. But some big questions remain: what about the time before we had weapons, and why do we still bother with these rituals now? And what about our other greetings, such as double-kissing, fist-bumping, cheek-slapping, and the like? How did they come about, evolve and spread? As we've seen, no one has yet come across a society that doesn't have some way of saying hello. Could it be that this gets to a deeper origin that transcends our differences? In worrying about what type of greeting to go for, are we missing their fundamental purpose?

As a historian, I mostly relied on written evidence to recon-struct and explain the past. Working on the early 1970s, I was often told that 'that's not even history'; yet the furthest we can go back using any kind of recorded word is still only around 5–6,000 years. And, even then, the earliest known accounts of any kind of greeting ritual don't come for another few millennia, with the ancient Greeks. In Homer's poem the *Iliad*, one of the oldest literary sources, set during the Trojan War and written sometime around the eighth century BC, the author describes a number of handshakes, though they were generally used to pledge allegiance and provide comfort. The handshake was also a common feature in early Greek funerary art, with art historians suggesting that it was most likely used as a symbol of a final farewell or reunion in the afterlife – though it also made a con-venient composition.[1] But perhaps the earliest representation can be found on a ninth-century-BC stone relief from Mesopotamia, in today's northern Iraq, showing the Assyrian king Shalmaneser III shaking hands with a Babylonian ruler to seal their alliance. Whatever the original function of the handshake, it's suggested that the Greeks adopted and popularised the custom following their excursions into Mesopotamia and then passed it on to the Romans, who in turn spread it to the rest of Europe via their empire. And so it has followed the fortunes of various conquests and encounters ever since.

But, notwithstanding the fact that the handshake exists in societies that could never have come into contact with the various European civilisations (tribes from deep in the Amazon, for instance), relying on the written and artistic record drastically limits our view. Furthermore, unlike our physical form, the behaviour of our earliest ancestors didn't leave any direct evidence: no bones or fossils to dig up that might allow us to follow its evolution. In reality, the first handshake or kiss would have gone without a trace. So, in the absence of physical evidence, how can we go back beyond the written record to see what the first greeting looked like and understand what it meant? First, we can look at the rest of the animal world, particularly species closest to our own. And second, we can look again at the various hunter-gatherer societies that might open up a window into our ancient past. As we'll see, neither approach comes without controversy, but they offer the best hope of looking back and getting to the core of what our greetings are really about.

For many, the notion that we can look for the roots of our behaviour in the animal world goes against what it means to be human. Bit by bit, though, scientific enquiry has challenged our illusions of specialness. The biggest blow came with the publication of Charles Darwin's *On the Origin of Species* in 1859, which suggested that all life evolved from something simpler and according to the same laws of natural selection. Fearing ostracism, even execution, it took Darwin another decade to publish his follow-up book, *The Descent of Man* (1871), in which he applied his ideas to mankind, suggesting that we're just another species of ape. Darwin was unwavering: it could no longer be argued that we were a separate act of creation – any differences were of degree rather than kind.[2] In more recent years, geneticists have proved what Darwin suspected, showing that we share around 98 per cent of our DNA with chimpanzees. Evolutionarily speaking, we really are no more than the latest ape, albeit with less hair and smartphones.

And so it was that I arranged to meet with the president of the Primate Society of Great Britain, Professor Simon Bearder, to discuss our connections with the rest of the animal world. Founded in 1967, the society was set up to promote research into all aspects of primate biology and behaviour. Given that the other primates are our closest relatives, much of its work helps bridge the gap between us and the rest of the animal world. For his part, Bearder specialised in the study of galagos, or bushbabies as they're commonly known. Living in the forests of Africa, these bug-eyed creatures look like a cross between a Disney rat and a raccoon. While they don't grab our attention in the same way as gorillas or chimpanzees, as Bearder explained, the interesting thing about bushbabies is that they are among the oldest primates, connecting the higher primates (including us) to the rest of the mammal world. With their powerful night vision and acute hearing, they are largely nocturnal, which is how most mammals would have lived when the dinosaurs were about. Having spent nearly half a century studying them, Bearder clearly had great affection for galagos, or 'my bushbabies', as he called them, learning to differentiate between all of their baby-like calls.

But the point of Bearder's research was not to fall in love with his bushbabies, but to understand them better and make sense of their behaviour. Here, like a lot of zoologists, he was influenced by the Dutch biologist Nikolaas Tinbergen, who in the 1930s was one of the pioneers of modern ethology. In many ways, Tinbergen had the mind of a five-year-old, driven by a simple nagging question: why? Why did his sticklebacks fight their reflection? Why do birds sing? Why do cats rip up carpets? The great achievement of Tinbergen was to get beyond observing these patterns of behaviour and to explain how and why they came about. At the time, the dominant view was 'behaviourism' – the theory that behaviour was essentially a reaction to external stimuli, famously shown by Pavlov's salivating dogs, and that any patterns were the result of experience rather than anything innate. But drawing on his years observing everything from sticklebacks to seagulls, Tinbergen showed how animal behaviour was more complex than a string of reflexes. It was also a matter of internal factors – what was fixed and instinctive.[3] To separate innate behaviour from what was learned, Tinbergen took newborn animals and raised them in isolation. Many such experiments have shown how these outcasts still develop the behaviours of their species. For example, when nestling song sparrows and swamp sparrows were raised side by side in a laboratory where they could hear tape recordings of both species' songs, each bird grew up singing only the song of its own kind.[4] Albeit unintentionally, a similar experiment was carried out at home with our cats, Alice and Bedford. They were born under my mum's bed and, sadly, orphaned a few days later when their own mum was run over. The vet said they had no chance, but my mum bundled them into a box and took them to work each day, where, with the help of colleagues, she fed them with pipettes. Even though they had no exposure to other cats and didn't receive any special training, both went on to develop typical cat behaviour, catching mice, ripping up the carpet and always digging a hole to do their business. As

Tinbergen suggested, certain behaviour is hard-wired, what he called 'innate fixed action patterns'.[5]

Moreover, while animals react to external stimuli, Tinbergen found that different species are preconditioned to react to different stimuli. Herring gull chicks, for example, peck at the red dot on their parents' bill, encouraging them to bring up food. In fact, Tinbergen discovered that they even peck at a red dot if it's painted on a cardboard cut-out or pencil. Similarly, just the sight of a bunch of red feathers is enough to get a male robin in a frenzy, as it defends its territory. Tinbergen also developed the idea of 'displacement activities', whereby a build-up of energy or motivation is released through seemingly irrelevant behaviour. For instance, when male sticklebacks meet at a territorial boundary, they dig madly, redirecting their aggression. Likewise, ducks will nervously preen themselves when courting a mate.[6]

As Bearder explained, though, perhaps Tinbergen's most important contribution to understanding animal behaviour was to ask how it relates to their survival and reproduction, which, he argued, was the ultimate question and explanation. Just like an eye or wing, Tinbergen contended that behaviour patterns evolved according to the same laws of natural selection. In this light, animal behaviour was never random or inexplicable. We might marvel at birdsong, the jamming of a male blackbird or the upbeat call of a cuckoo, but they're not singing for pleasure. They are busy protecting their territory and attracting mates.

Taking this perspective to another level, it was one of Tinbergen's students, Richard Dawkins, who in the early 1970s popularised the 'gene's eye' view of the world. In the great struggle for survival and reproduction, Dawkins argued that it's genes (short sections of DNA that contain the instructions for the physical characteristics of every living organism) that are the basic unit of natural selection. It's genes that are the true replicators; evolution and reproduction are merely the process by which they become more or less numerous. As for all the animals (including humans), they're mere 'survival machines', expendable containers, blindly programmed by genes in their quest for immortality.[7] Like the body, behaviour can be seen as an extension of the same selfish ends, simply helping the genes to survive and reproduce, whether through building a nest or courting a mate.

For Bearder, who himself was trained by another of Tinbergen's students, all of this was invigorating stuff. It gave him a way of understanding his bushbabies. 'The more I looked at them,' he said, 'the more I saw them getting on in a matter-of-fact way.' His face lit up: 'They haven't, but it's as if they've read a textbook. If I ever learned anything,' he added, 'it's that the males will always make sure that they're in the right place at the right time.'

The big question was how much all of this applied to our own species. As Bearder began his doctoral studies back in the 1960s, biologists were beginning to look harder at humans. The fundamental question was as Tinbergen put it: how much of our behaviour is simply a matter of survival and reproduction?

'Basically, what I see is animals and people trying to have sex,' Bearder said, looking up from his sandwich. 'It's that that's driving nearly everything.'

While it may be simplistic, it is hard to resist the conclusion. After all, it doesn't take a degree in zoology to see how much of our efforts are geared towards finding and keeping a partner. Yet, despite stating much of the obvious, turning the zoologists' lens onto humans created a storm in academia, pitching colleagues and entire departments against each other. Taking the lead from the likes of Franz Boas and his students, many in the social sciences vehemently rejected the notion that human behaviour was in any way a matter of instinct or innate urges. As the Spanish philosopher José Ortega y Gasset put it: 'Man has no nature.' Instead, it was believed that context and culture determined how we behave. Things came to a head in 1975, with the publication of *Sociobiology* by Edward Wilson, a Harvard biologist most noted for his work on ants. In the final chapter, he turned to humans, suggesting that we should look on ourselves as if we were zoologists from another planet, treating our own species like any other.[8] Following Tinbergen, he argued that we could even relate matters of our culture and social organisation back to the same evolutionary ends. As others went on to suggest, things like art and music were ultimately about social bonding and attracting a mate.

Coming at the end of the Vietnam War and during the civil-rights movement, Wilson's book fuelled what was already a highly charged atmosphere on university campuses. His whole approach was dismissed by many as Darwinian fundamentalism – the latest abuse of the great man's ideas. At best, it generated untestable 'just so' theories, not so different from Kipling's story about how the elephant got a long trunk. At worst, though, it helped to justify inequality and the oppression of women, undermining everything that Boas's students had fought for as they sought to break us free from this sort of biological determinism.

It was in this heated atmosphere that, on the other side of the Atlantic, Bearder was appointed Professor of Physical Anthropology at Oxford. At the time, he found that he too was treated with suspicion. Many of his colleagues couldn't understand what he was doing in a social-sciences department. The disconnect was mutual: 'I wondered how some of them had children,' he chuckled.

As the academics slugged it out, the ethologists' approach was gaining popularity, especially when it came to understanding our day-to-day interactions. Most famous at the time was Desmond Morris, another of Tinbergen's students, who in 1978 published *Manwatching: A Field Guide to Human Behaviour.* Morris believed that humans could be observed in the same way as birds. The city was just a human zoo. His book (now retitled *Peoplewatching*) went on to become a best-seller, helping pioneer the body-language profession. But, in reality, many of his insights went back to his old mentor. Borrowing the terminology, Morris characterised certain core behaviours as 'fixed action patterns' – preprogrammed responses to specific stimuli, just like in animals. A clear example is one of the first things we do when we come into the world: latch onto our mother's breast and suck.[9] We might also think of some of the elements of our greetings that we do unconsciously, which Kendon described – the eyebrow flash or even the smile, for instance. Morris also applied Tinbergen's notion of 'displacement actions' to humans. Whether twiddling hair, biting nails or jogging on the spot when waiting for someone at the airport, we're constantly doing them, releasing our pent-up energy. Even clapping can be seen in this way: originating as a way of channelling our extra enthusiasm – helping to explain the greeting I witnessed in Namibia (described in the Preface).

Just as ethologists were demonstrating how much we are like other animals, they were also narrowing the gap from the other direction, showing how much animals are like us. It was a century earlier that Darwin, having set out his theory of evolution, turned his pen to the minds of humans and other animals, identifying six core emotions: anger, happiness, sadness, disgust, fear and surprise. Applying the same principle of continuity that connected us physically to the rest of the animal world, he reasoned that other animals must experience similar emotions. After all, he only had to glance at his dog to know whether it was feeling happy or dejected.

In the intervening period, however, this tendency to see our own traits in other animals – what's known as anthropomorphising – was looked down upon, dismissed as the ultimate professional sin. With behaviourism dominant, the idea that animals had feelings and personalities became the stuff of Disney films rather than scientific enquiry. More generally, transposing the subjective experiences of humans onto animals was seen as an affront to objective study. There was no way of knowing what they felt (if, indeed, they felt anything).

But this criticism could be turned on its head: if Darwin was right about our physical continuity, surely it made no sense to keep emotions separate. The person who did most to take up Darwin's lead was the other great pioneer of modern ethology, the Austrian zoologist Konrad Lorenz. Like Tinbergen, whom he became great friends with, despite their ending up on opposite sides during the Second World War, Lorenz viewed animal behaviour as a complex interplay of innate reactions and responses to external stimuli. But whereas Tinbergen was the consummate fieldworker, maintaining a dispassionate eye, Lorenz was like a real-life Dr Dolittle, deliberately forming strong bonds with his animals and trying to imagine what was going in their minds. While he warned of the dangers of anthropomorphising, his most popular writing is full of it. At times, his 1949 book *King Solomon's Ring* really does read like a Disney story, as he describes how he grew up with his capuchin monkey Gloria, raised baby ducklings and befriended his greylag geese. But the star of the show was Jock, his jackdaw. Lorenz kept a whole colony of jackdaws on his roof in Vienna, but it was Jock whom he grew closest to, taking him for walks and bike rides. He even describes how Jock fell in love, casting 'glowing glances' into his partner's eyes.[10] Watching his jackdaws swoop in the wind, Lorenz was in no doubt that they were playing, flying for pleasure. But he also detailed a darker side, describing how animals experienced loss and pain. When one of his greylag geese lost its partner, it showed unmistakable signs of grief: sinking eyes and a sagging face.[11] To be sure, Lorenz conceded that he could never be certain about their inner feelings, but he was convinced that other animals were capable of experiencing emotions similar to our own.

So what of our closest relations? Despite our being so similar genetically to chimpanzees and sharing a common ancestor, the differences are obvious. Chimpanzees are covered in fur, walk on all fours and sleep in trees. They can't talk or send rockets into space. When Queen Victoria was introduced to one dressed in a sailor's outfit at London Zoo, she was appalled, describing it as 'frightful, and painfully and disagreeably human'.[12] But today some scientists feel that chimpanzees are so close to humans that they should be reclassified and included in our own genus, *Homo*. Taking the scientists' lead, the US lawyer Steven Wise has devoted his career to fighting for chimpanzees, along with the other great apes, to be recognised as 'legal persons' with the same rights as a child.

But the person who has done most to shift our perceptions is Jane Goodall, who in the 1960s became the first person to study chimpanzees in the wild. Growing up on a diet of Dr Dolittle and Tarzan books, she dreamed of going to Africa to study animals. Then, aged twenty-three, with most of what she'd learned coming from watching her dog, Rusty, she got her chance. During a trip to Kenya, she was introduced to the British palaeontologist Louis Leakey, who had become famous for tracing the origins of mankind to Africa. With the fossil record scarce, Leakey turned to the great apes, hoping they might tell him something about how early humans lived. He was immediately impressed by Goodall and employed her as his assistant, eventually sending her to Tanzania to study chimpanzees. And so it was that, in July 1960, Goodall found herself setting up camp in the Gombe River Reserve on the shores of Lake Tanganyika. Traipsing through the thick mountain forests, it was initially tough going, since the chimps ran off as soon as they saw her. But slowly they got used to Goodall and, with the help of the odd banana, she was able to get close enough to study their behaviour.

It was three months in that Goodall got the breakthrough that would catapult her to global fame. Early one morning, peering through the trees she noticed a male chimpanzee feeding at a termite mound. As she focused her binoculars, she observed something that no one had ever seen before: the chimpanzee broke off a stick, stripped its leaves and fished for termites. He had made a tool. Until that moment, tool-making was thought to be a unique ability that separated humans from animals. Goodall telegraphed Leakey, whose infamous reply didn't miss the significance: 'Now we must redefine "tool", redefine "man", or accept chimpanzees as humans.'[13] Little more than an amateur, Goodall had made one of the most important scientific discoveries of the century.

Today, aged eighty-two, Goodall spends most of her time trying to protect chimpanzees, travelling over 300 days a year to promote her various causes. Often appearing with her toy chimp, Mr H, she speaks with the weight of the world in her voice, warning about the dangers of deforestation and overpopulation. But there's also a mix of hope and wonder in her tone, as she targets younger generations and recounts her early experiences.

Unsurprisingly, it took me weeks of pestering to convince her assistant that I was worth talking to, in between her efforts to save the planet. I have to admit that I was feeling nervous as I waited for Goodall to call; after all, it's not often that you get the chance to speak to someone who has redefined our species or been parodied in *The Simpsons*. She was on a coach to Heathrow to catch a flight to Iceland, before heading on to Kenya, Uganda and, finally, back to Tanzania, where it all began nearly sixty years ago. My first question was so garbled that I couldn't understand it myself. But Goodall was kind enough to say that she thought the line was breaking up before asking me to repeat it. Thankfully, her own response was crystal clear: 'They're unbelievably like us.'

In many ways, this summarised her findings from what turned out to be twenty-five years of research. At times, her resulting book, *The Chimpanzees of Gombe*, which also earned her a PhD, reads like a *Watching the English* for the entire species, as she describes the daily lives of David Greybeard (the original termite fisher), Gremlin, Passion, Leakey and the rest of the 150 chimpanzees she got to know. What really fascinated Goodall was not their tool-making abilities, but their social lives. Back home, scientists criticised her for giving her subjects names and personalities. But, as she explained, it wasn't that she gave them personalities: 'I was merely describing what I saw.'[14] Spending so long with the chimpanzees, Goodall was able to watch individuals grow up and change. Like us, they have an extended childhood and go through adolescence. Their society is complex and shifting, built on a range of relationships.

While chimpanzees spend much of their day searching for food, there is still plenty of time for hanging out. By far the most popular activity is grooming – carefully combing through each other's hair and picking out any bugs or nasty bits. It's essential for keeping them free of parasites. But it's also relaxing, calms tensions and helps to build relationships. Chimpanzees can spend up to three hours at a time grooming each other, often in big groups, sprawled out like they're in a jacuzzi. 'Social grooming', as the experts call it, is chimpanzee bonding time. To keep themselves amused, they'll also play games, roll around and wrestle. They try new tricks, and tickle and tease each other.[15]

I asked if she'd ever seen a chimpanzee show signs of embarrassment, bringing the conversation closer to the matters at hand, wondering if it's a uniquely human affliction – maybe this was where we could draw the line. She couldn't be sure, but described a moment when a young chimp, Freud, climbed to the top of a plantain and began swinging about, with his mother and uncle, Figan, below. 'It was like he was showing off to Figan, who he absolutely worshipped,' she said. But then, as he built up speed, the stem broke, sending him tumbling into the grass. 'A few seconds later I saw his little head poking up, checking if anyone had seen,' Goodall laughed, as if it was just the previous week.

But for all that Goodall was charmed by the chimpanzees, she also uncovered a darker side.

'What happens when chimpanzees from different groups meet?' I asked.

'Oh, they try to kill each other. Unless there's a female about.'

This came as a shock to Goodall, who, like many at the time, had assumed that chimpanzees live together harmoniously.

But the truth is that when they meet rival groups the results can be fatal.

All in all, it was clear that studying our closest relatives provided a window into some of our own behaviour and its origins. With Goodall paving the way, primatologists have built an ever more detailed understanding of chimpanzees, continually narrowing the gap between us and them. As she puts it: 'The line is wuzzy and getting wuzzier all the time.'

So what about greetings? During her time at Gombe, Goodall observed a number of behaviours that chimpanzees perform when reunited. Much like us, they live in fission–fusion communities, whereby individuals and groups come and go, so there were plenty of greetings for Goodall to see. And much like us, the nature of the greeting depends on the relationship and how long the chimpanzees have been apart. Sometimes they'll only manage a glance and grunt. But if they are close or have been separated for a long time, they can get quite intimate. What's striking, if not surprising, is how similar their greetings are to our own. One of the most common is the hug. Unlike some of our own efforts, though, chimpanzees fully wrap their long arms around each other, sometimes burying their heads in their partner's chest. There's a clip on YouTube of Goodall meeting Wounda, a rescued chimp, in which she is swallowed by a big hug, reducing many who have watched it to tears.[16] It's perhaps no coincidence that Goodall herself has become renowned for the quality of her hugs.

To initiate contact, chimpanzees often offer and touch hands – a sort of handshake but without the shaking part. They also kiss each other on the lips, sometimes for a minute or more. To counter accusations of anthropomorphising, some primatologists prefer to describe it as 'mouth-to-mouth contact', though this seems a bit like calling a spade an 'earth-moving utensil'. However you describe it, chimpanzees kiss. Goodall even tells the story of how when a female arrived in a group she rushed up to the alpha male and held out her hand. Almost regally, he took it to his lips and kissed it.[17]

As Goodall often demonstrates at the beginning of her talks, there's also a vocal element to chimpanzee greetings. To her audience's delight, she starts with a soft pant, which builds in speed and volume until it becomes an excited hoot.

'What do you think they mean?' I asked, resisting the temptation to ask for an impression.

'Oh, you know, "I'm so pleased to see you" or something like that.'

Reading Goodall's book and talking to her, I felt like I'd cracked it: I'd found the origins of our most common greetings. Far from being a few thousand years old, they go back to the dawn of humanity, most likely to a common ancestor. To talk of the Mesopotamians, Greeks or anyone else inventing the handshake misses the point. Watching our closest relatives, it's clear that we've always been doing them in some shape or form.

But having dug into the history, I hoped to go further and asked Goodall what she thought the purpose of these rituals was and how they came about. A light bulb went off as she gave her answer: 'Well, I think that a lot of chimpanzee behaviour, like our own, goes back to the relationship they have with their mothers.' It was so obvious. I thought back to the moment at Heathrow when the girl jumped into her mum's arms, wrapping her arms and legs around her. Aside from the fact that she was a bit big, it had seemed so natural. As Goodall explained, chimpanzees are entirely dependent on their mothers for the first four years of life. And, in mammals, this goes back even further, to our time in the womb, where we're kept nourished and secure. It's no surprise that our first experience of the big bright world is one of sheer terror. As Morris says, it's our mother's embrace that calms us, bringing us closer to her heartbeat and where we came from.[18]

Among chimpanzees, the embrace also serves a more practical function over the subsequent days and years – it's how they get about. Unlike a newborn lamb or fish, which gets going straight away, infant chimps have to cling on to their mothers. The same is true for us; only we don't have any fur to hold on to so have to hang on extra tight. And when distressed, it's our mother's embrace that comforts and protects us. Excuse the ugly word, but each time we hug it's as if we become enwombed.

Similarly, clasping hands has a basic function, acting as a natural leash. When visiting his old friends, the group of chimpanzees at Arnhem Zoo, the Dutch primatologist Frans de Waal found that the matriarch, Mama, always greeted him by extending her hand, which was the signal she used to get her offspring to jump on her back.[19] Reflecting a different impulse, reaching out is also what babies do when they want something and how mothers pass food.

The kiss is less obvious but, looking at chimpanzees, a likely theory is that it stems from the practice of mouth-to-mouth feeding, whereby mothers chew up food and pass it directly to their offspring. You can even watch a demonstration in a video posted by the actress Alicia Silverstone showing her feeding her ten-month-old son.[20] While most people reacted with disgust, it's likely that Silverstone was merely going back to the days before blenders and baby food. It's even more common in other species, particularly birds, which use their beaks to pass insects or bits of mouse into their chicks' gaping mouths. Some such as seagulls and penguins go further and regurgitate food, literally vomiting into their offspring's mouths, while the chicks will nuzzle at their parents' beaks for more. This feeding method is later reflected in the behaviour of courting birds with their own version of kissing: billing. While writing this chapter, I sat outside watching a couple of wood pigeons on our roof rubbing beaks, complete with the setting sun behind. There's no doubt that they were having a good smooch.

Alternatively, it's been suggested that kissing comes from suckling. Whether we kiss on the lips or blow one across the room, we make a similar sucking action to the one we used to get milk from our mother's breasts. Studying baboons, scientists at the University of Chicago concluded that the lip-smacking gesture, which they often use as a greeting, also comes from this same sucking pattern.[21] Taking a more literal approach, according to a visitor from the Austrian embassy in Indonesia, on entering a village in Papua New Guinea acquaintances were required to take a brief welcoming suck on the breast of the chief's wife.[22]

Whatever the exact origin of kissing, taken together, our most common greetings reflect the fact that we've never grown up. They're what Irenäus Eibl-Eibesfeldt calls 'infantilisms' – baby behaviour.[23] Aside from our greetings, we use them on all sorts of occasions, from when we've got something to celebrate or commiserate to when we want to express gratitude or give reassurance. Watching the chimpanzees at Arnhem, de Waal found that, as well as greetings, kisses are most commonly used after a fight as a means of reconciliation. Like us, chimpanzees kiss and make up; it's only then that their relationships resume.[24] We also revert to this kind of baby behaviour with our partners, cuddling and kissing. And suddenly passing food mouth-to-mouth can seem like a good idea (at least, so the *Kama Sutra* guide tells us). New couples also often revert to a kind of baby talk, calling each other things like baby, cutie pie and kitten, while adopting the tone of a four-year-old.

———

As Goodall's bus pulled into Heathrow, I really felt like I'd got to the bottom of our greetings – where they come from and why. But as she headed off to talk about the plight of chimpanzees and save the planet, suddenly it seemed like trivial stuff. As well as her various education initiatives, Goodall described how her team is helping to establish corridors across Africa in an effort to link different chimpanzee communities. But, as she explained, it's a depressing situation: their population has fallen from around a million at the turn of the last century to fewer than 300,000 today, mostly due to deforestation and the bushmeat trade.

'We're destroying the planet, Andy,' she said.

It was tough to know how to respond.

'Well, thanks very much for taking the time to talk and good luck with it all,' I said.

Once I'd got over the feeling of inadequacy, my conversation with Goodall left me more certain of what I already knew: that the roots of our own behaviour can be found in the animal world. But before we get carried away with seeing chimpanzees as a mirror of ourselves and make deductions about human nature, perhaps suggesting that male dominance and aggression are simply an extension of our primate heritage, we should consider another species, one that, while less known, is as closely related to us: the bonobo. Only classified as a separate species in 1929, the physical differences between chimpanzees and bonobos are marginal: they have a slighter build, a flatter and darker face, and pink lips – a more intellectual-looking chimpanzee. But the contrast in behaviour is striking. Living on the other side of the Congo River, which most likely caused their separation around 1.5 million years ago, bonobos are often said to have built a hippy paradise. Rather than being headed by ultra-competitive males, their society is led by the females, who form tight and lasting bonds. Fights are uncommon, but when they do break out they are quickly resolved with one thing: sex. In fact, sex is used for just about everything, from making up to saying hello. And, significantly, much of it is between females. Even when meeting for the first time, they will wrap themselves around each other and rub clitorises, squealing loudly. From their big grins, it's clear they're enjoying themselves, but their sexual appetite isn't about satisfying some hedonistic drive. For bonobos, sex is about bonding – it's what keeps the sisterhood so tight.[25]

Even though some would prefer to keep bonobos out of public view, when it comes to looking for connections to the rest of the animal world, we should give them as much attention as chimpanzees. In fact, the two species have been used as short-hand for what's best (bonobos) and worst (chimpanzees) about human nature, with de Waal suggesting that we're bipolar apes.[26] But despite the efforts of various countercultural movements, humans have historically been more like chimpanzees, with public life dominated by men and conflict common. And, when it comes to greetings, we generally stop short of sex.

But even if we accept that we are closer to chimpanzees, some big questions remain. It might be clear that our most common greetings reflect the mother–child relationship, but how did they become ritualised and why do we perform them with just about anyone? And what about the greetings that bear no resemblance to baby behaviour: the high five, fist bump and face slap, for example?

It's here that it's useful to look again at some hunter-gatherer tribes. For the vast majority of our 200,000 years, this is how modern humans have lived: roaming around in nomadic bands, searching for food – essentially on lifelong camping trips. While most of us now live in a way that bears little resemblance to the lifestyle of our ancient ancestors, the remaining hunter-gatherer societies, such as the San of southern Africa or the Pirahã of the Amazon, offer a potential window into our common evolutionary past.

Evolutionary psychologists look for commonalities across cultures, particularly those shared with hunter-gatherers. Any universal aspects of behaviour, they argue, are likely to be adaptations, selected for their survival value, just like an eye or a hand. Starting out as a sceptic, the anthropologist Donald Brown ended up compiling a list of over 600 traits, even describing what he called the 'Universal People'.[27] It turns out that they (we) share everything from biological functions, such as pain and sexual pleasure, and basic inventions, such as fire and shelter, to deeper psychological needs, such as music, folklore and religion. Universal People get married, live in families and mourn the dead. They tell jokes, trade, fight, experience sexual jealously, make laws and have a sense of fairness.

Greetings are also on Brown's list, but here even Darwin would have been sceptical. While he observed that certain expressions are universal, he recognised that our gestures vary from culture to culture, that they're learned and passed on between generations – though he did wonder about the shrug. But when it comes to greetings, his own account of his voyage on HMS

Beagle suggests that the evolutionary psychologists have a case. Arriving on a remote island off the coast of South America on 18 December 1832, Darwin, then a fresh-faced graduate, had his first encounter with one of the local hunter-gatherer tribesmen, who gave him three hard slaps to his chest and back before 'baring his bosom'. Without hesitation, Darwin returned the gesture, much to the delight of his new friend.[28] To adapt one of his own phrases, here was a difference in form rather than kind.

So, to go back to Tinbergen's big question, could it be that our greetings are an adaptation, having some sort of reproductive or survival function? In many ways, I think that the answer is yes and that, even today, our rituals bear the imprint of this evolutionary logic. In fact, we could even say that we have a 'greetings instinct'. In short, our greetings both express and help to manage two sides of humanity that have come to define us: our capacities for cooperation and competition. This inherent duality has been expressed in many ways – compassion and aggression, friend and foe, us and them, love and hate, even good and evil – but they all get to the same basic tension that conditions so much of how we feel and behave towards each other.

Philosophers have long argued the toss, but it's perhaps our cooperation and sociability that have characterised our species most or at least determined our successes. Of course, we're not alone. In fact, at the most basic level, every complex living thing (every elephant, fish or tree) is a triumph of cooperation between cells – a process that began some 6–800 million years ago, when single cells started to combine to produce multicellular organisms and then the first animal, most likely some kind of jellyfish. Then, around 565 million years ago and 2 kilometres under the sea, cooperation was taken to a whole new level when the first organisms began to reproduce by having sex.

And sex necessitated cooperation. Gradually, species evolved, especially birds and mammals, in which the offspring depended on parental care, creating another level of dependence. As Bearder concluded, these two relationships – between sexual partners and between mother and child – became the basic units of sociality.

Whether in packs, herds, shoals, flocks or entire colonies, members of the same species came together in bigger groups to mate, hunt and protect themselves. For our part, we have taken cooperation and social living to higher levels, living and working together on an unprecedented scale, forming institutions, cities and whole civilisations. For good reason, we call ourselves 'supercooperators' and 'hyper-social'.[29] In fact, evolutionary psychologists have even argued that it's our sociability that drove the thing that sets us apart most: our supersized brains. While we tend to associate our grey matter with intelligence (our unrivalled ability to solve problems), Robin Dunbar at the University of Oxford has recently developed the hypothesis that our brain growth was first and foremost a function of living in bigger groups and all the advantages that come with this. So, extrapolating from chimpanzees, which live in communities of around fifty, we can expect humans, having brains that are three times larger, to live in communities of around 150. Or, put slightly differently, 150 is the number of people we can have meaningful relationships with at any one time. Broadly, looking across the historical and anthropological record, whether it's the average size of Neolithic villages, the size of companies in the military or the number of Christmas cards we send, this seems to be the case, with 150 (what's become known as 'Dunbar's number') being strikingly common when it comes to human social organisation.[30] And, as we'll see in the next chapter, the part of the brain that has grown most over the last couple of million years is precisely the bit that helps us get along.

Moreover, we also go beyond cooperation, being capable of great acts of kindness, such as giving to charities and jumping into rivers to save each other – acts that appear to go against our own interests. But before we go redrawing the line, perhaps attributing all this to our various moral and religious teachings, we should look again at the rest of the animal world. In *The Family of Chimps*, a film about the group of chimpanzees at Arnhem Zoo, for example, there is a scene in which, after a number of attempts, a younger member, Walter, manages to climb a tree protected with electric wire. He first takes a few handfuls of leaves for himself, but then starts throwing them down for the others.[31]

There's always a chance that we will take advantage of all this goodwill, but even for evolutionary biologists it has a certain logic. After all, the main theme of Dawkins's *The Selfish Gene* is explaining all the altruism in the world. There's the theory of 'kin selection' – the idea that we're prepared to make sacrifices for people who are genetically related to us, as a way of promoting our own genes' survival (something that goes some way to explaining the descending order of presents at Christmas). Or there's 'reciprocal altruism', otherwise known as 'you scratch my back and I'll scratch yours'. So, just like other animals, we perform a favour (share our food or lend a hand) on the understanding that we'll be paid back later, when we're in similar need. Sometimes, these acts can get extreme, even competitive. So it's not uncommon for certain birds to fight to take the most vulnerable spot on a roost or for us to try to outbid each other at charity auctions and on fund-raising pages. It may be a worthy cause, but we may have a more selfish motive: to boost our social prestige and status.[32]

But while our altruism can end up looking like a sleety mix, there are still acts that can't easily be explained by any evolutionary model. I often think of my time in Sudan, during my journey to Cape Town. I found myself lost in one of Khartoum's crazy bus stations, unsure which of the hundreds of buses to get on. Just as I was about to give up and walk, a man stopped to check if I was OK, put me on the right bus and paid for my ticket, refusing to accept any money. I could recount dozens of these moments, from both my African trip and time as a diplomat there. The Sudanese are famous for it. But others tell of similar experiences travelling in the likes of Iran, Syria and Afghanistan. In fact, it's as if there's an inverse correlation between the Foreign Office travel advice and the generosity of a people (but that's a different book). But, again, it's not just us. Chimpanzees have been known to adopt unrelated babies and care for the sick. Or there's the herd of elephants in Kenya that kept stopping for their injured friend, despite him being no use to them.[33]

In the end, for all our moralising and whatever the genetic calculus, part of what motivates our cooperation and altruism surely comes down to a simple fact: it can make us feel good. There are people who love to give presents and bake cakes, not because they expect something in return or are trying to boost their status, but simply because it gives them pleasure to see pleasure in others. What's more, we seem to be primed this way. It makes sense, as Darwin recognised, that nature has endowed us with a set of emotions that motivate us to get on with each other.[34] It's the same with the pleasure we get from eating or sex. And it helps to explain why soldiers and rock climbers, people whose lives depend on cooperation, feel such strong bonds. Conversely, when we cheat or exploit each other, we feel a range of negative emotions, such as guilt or shame. In part, we are moral *because* of nature, not despite it.[35]

So, if we are to talk of any greetings instinct, it is first and foremost a function of our 'social instinct'. Above all, we are signalling our desire to get on. But the other side of all this is that there would be no need for these sorts of visual and vocal rituals if there wasn't also the potential for competition – for enmity, hate and aggression. You only have to look at a kids' playground or flick through a picture book of the twentieth century to know that humans can a be a nasty bunch, doing anything to get our own way, including killing each other. Perhaps most chilling, though, was Goodall's discovery that our closest relatives are not the peace-loving creatures we once thought, but can also be competitive and violent. In 1974, she observed the beginning of what she called the 'four-year war' between two chimpanzee communities at Gombe. Originally, they had been one, but split when the number of males became too large. Once contact had broken off, members of the original group, which outnumbered the breakaway community, began to conduct a series of violent raids. Goodall was shocked by what she saw, as the males formed gangs and carried out lethal attacks until they'd killed or chased away the entire group.[36] It's been suggested that Goodall inadvertently fuelled the conflict by introducing feeding stations, but a number of studies have since shown that it doesn't take extra bananas to get chimp communities fighting.

Looking at our own past and across the animal world, it's hard to resist Lorenz's conclusion that the potential for aggression is innate. Taking Darwin's theory to its grim conclusion, in a world of finite resources, only those who can compete will make it. Many species even have inbuilt weapons – horns, fangs and claws are not all about catching prey. Overwhelmingly, though, it's just one half of the animal world that's been responsible for all of the violence. And, in the long run, it always comes down to a struggle for resources, whether food, fuel or mates. Beyond all of the wars, we're constantly battling it out, whether on the playground, sports field or dance floor, in business or in politics. We seem forever prone to see the world in terms of those who are inside and those who are outside our group, often with fatal consequences.

Ending on an optimistic note, the final photo in my twentieth-century picture book perhaps best captures the essence of the human spirit. It shows a scene from a modern take on Beethoven's opera *Fidelio* in which the main character, Leonora, persuades a jailer to let his prisoners out of their cells for some light and fresh air. The central theme of the opera is simple: that love conquers hate. Put differently, perhaps it's that we're left in a state of cooperative competition in which self-interest is best served by the common good. To be sure, competition came first, when life was all about individual cells, but when it comes to debating human nature – whether we're inherently competitive or cooperative, more chimpanzee or more bonobo – the answer is clearly that we're both, whichever side the philosophers and scientists come down on. The two come together, side by side, inhabiting the same mind. Like two siblings playing, we can go from being full of love and laughter one minute to being consumed by tears and hate the next. We really are bipolar apes.

If our relationships are shaped by these twin impulses of cooperation and competition (or however else you want to put it), greetings can be seen as playing a vital role, coming at the key moment of reunion or when we first meet someone. They're the frontline of our social encounters. In a world where cooperation and competition cohabit, it's through these opening rituals that we signal – and sense – our intentions towards each other. To this end, our greetings have several functions.

First, and most obviously, they reaffirm our ties. Like chimpanzees, we live in fission–fusion communities where people come and go and our relationships constantly need updating. In our reunions, we tend to revert to behaviour that characterised our closest bond of all. We hug, kiss, hold hands or, if we're lions, rub noses and lick each other just as we did with our mothers (answering my own mum's question, it's why Alice nuzzles her each morning). The closer we are and the longer we've been apart, the more intense the greeting. If we've been separated for a particularly long time, we may even give something to renew our ties. Many animals often bring bits of food or nest material when reuniting with their mates. One of the most striking examples is the food pass between hen harriers. The silver male 'ghost bird' flies above his mate and calls, dropping his prey, while she swoops underneath and catches it mid-air.

These rituals can be regarded as largely symbolic: we're not supposed to get any sensory pleasure from them. But looking across the rest of the animal world, it's clear that physical greetings do not just express bonds: they are instrumental in making and maintaining them. For example, Lorenz found that the 'triumph ceremony' of his greylag geese was what kept them together.[37] And there's no doubting that bonobos get pleasure from genital-rubbing. Perhaps the most unusual example, though, comes from seahorses, which form monogamous pairs. Each morning they locate their mates and greet each other by changing colour and dancing in circles, usually for six minutes or more. Researchers at the University of Cambridge concluded that the ritual was key to maintaining their bond.[38] Our own rituals might not be as elaborate, but if we don't do them, perhaps because we're not sure which to go for, we can feel that something is missing.

Among many species, the greeting merges into some form of grooming, which acts as an extra demonstration of friendship. We might lack the fur, but we often look each other up and down, maybe brushing off a crumb or offering to take the other person's coat or suitcase. These are small acts but, in the absence of grooming, they play a role in reaffirming our bonds. We then ask the universal question, 'How are you?' Again, we're not simply going through the motions, conforming to what's culturally expected; when we enquire after someone's well-being, we are demonstrating a fundamental aspect of what it means to be human: empathy. Given that we rely so heavily on each other, it's no surprise that we are endowed with the capacity and inclination to understand and feel what others are experiencing. The extra small talk, what's often regarded as meaningless chat, can be seen as a form of social grooming.

In fact, Dunbar has argued that this was the origin and purpose of language, most likely evolving from our calls and gestures. As he observes, just like other animals, our relationships need to be serviced. But whereas animals spend hours grooming each other, we don't have the time (let alone the fur) to go around our extended social networks and service everyone. Instead, we developed language as a means of connecting with each other, allowing us to service more than one person at a time. In general, we can have a proper conversation with up to three other people (any more than that and it starts to become a series of mini performances or to break down into smaller units), which fits with Dunbar's thesis about the size of our brains and social networks being three times the size of chimpanzees'. If all this sounds far-fetched, think of what we spend most of our time talking about: we're not exchanging important information, but asking what's on each other's minds, discussing relationships, sharing stories and making each other laugh – what we otherwise call gossip. Even when we've got nothing much to say or don't know someone, we still find something to talk about, whether it's the weather, our running times or what's on telly – we're servicing our relationships. Conversely, silence can feel uncomfortable. Listening in to the conversations of his academic colleagues over lunch, Dunbar found that even what they talked about could be mostly classed as gossip. There's no way of knowing for sure, but there's good reason to think that the first words that came out of our mouths were some kind of social grooming – maybe even a version of 'Hi, how are you?'[39]

Our encounters with strangers or people we're uncertain about can be particularly tense moments. Here, greetings provide a quick signal of our intentions – what Eibl-Eibesfeldt calls 'appeasement rituals' or 'pacifiers'. It helps to explain why we associate our most common greetings – the wave and handshake – with showing that we have no weapons. Similarly, Maasai warriors stick their spears in the ground, while the twenty-one-gun salute comes from the practice of demonstrating that all the guns on a ship had been discharged.

In uncertain situations, gift-giving becomes a form of peace offering. It perhaps explains an unforgettable moment for Goodall when David Greybeard (the termite fisher) approached her and placed a palm nut in her hand. It's a practice that was used by the likes of Captain Cook and other explorers when they encountered different communities on their voyages. Anthropologists still rely on it today when trying to make contact with isolated communities. Something similar was going on when I found myself sitting in a *ger* in the middle of Mongolia, having been invited in during a torrential storm. As I sat by the stove, the owner offered me a cup of fermented yak's milk (which tastes like blood) and sheep's testicles (the local delicacy). Despite being repelled by the thought, deep down, I knew I shouldn't refuse, not so much because I'd be insulting their culture, but because it would be refusing their offer of friendship.

The most universal and effective appeasement ritual is also the simplest: the smile. The right kind of smile can be like switching on a light, immediately breaking the tension – though it's perhaps surprising that the baring of teeth should be reassuring. In fact, it's likely that it evolved from something more threatening: the snarl. When chimpanzees roll around, they make a play face, consisting of an open mouth, exposing their lower teeth. But as soon as it turns nasty they reveal both rows of teeth, as if symbolising just how close love and hate can be. Similarly, mandrills – the world's largest and most colourful monkeys, and which have a complex system of gestures – reveal their dagger-like teeth as a sign of conciliation, usually when a more dominant member of the group arrives. However, the yawning version sends an altogether more aggressive message. Eye contact is the same, teetering along a fine line. Like many primates, mandrills use the stare to intimidate. And so it is with us. A glance can break the ice, but if held for too long can quickly become threatening. We pacify by averting our gaze.

While laughter might not be a greeting in itself, it often performs a similar function in our opening exchanges. We often blurt out a laugh during tense encounters, such as at a job interview or on a first date. It's not that anything's funny – far from it – but rather that laughing releases endorphins and breaks the tension. It's why we English often inject some sort of joke into our opening exchanges. It turns out that it's not so different with the most famous laughers of all – hyenas – which tend to giggle when they are threatened or nervous, such as when being picked on by other members of the clan. For us, the great thing about laughter is that it's contagious. As the neuroscientist Robert Provine concludes, 'laughter is more about relationships than humour.'[40] Whenever I walk past the local pub landlord's dad shuffling along with his old greyhound, he always jokes, 'Just going out for a quick ten miles.' And I always laugh. It's why our clumsiness can work to our benefit. When we mess up our greetings, the element of slapstick makes us laugh, easing the tension.

As we go about our day-to-day interactions, it might seem far-fetched to interpret our greetings as a form of appeasement, but consider the difference in the way we behave when we're in the countryside and the city, captured by the scene in *Crocodile Dundee* I mentioned in Chapter 5. Why is it that we suddenly find ourselves saying hello or passing a smile to everyone when we're out in the sticks? As suggested earlier, it could just come down to the fact that we can't say hello to everyone in the city, since life would become unmanageable. Or perhaps we feel more relaxed when we're away from all the traffic and breathing in fresh air. Maybe. But there's something else going on. In the city, we're part of the anonymous herd and, conversely, feel some protection from it. But when we're walking alone in the fields with no one else around, strangers become more threatening. So we go beyond civil inattention and pacify each other with an extra smile or hello. It's the same entering a lift. If it's full, averting our gaze is enough. But if there's only one person inside we'll feel the urge to make extra contact. And, if we don't, we squeeze the railings extra tight.

Lorenz went further, suggesting that greeting rituals are a form of 'redirected aggression'. But while his starting point was that aggression is innate, he was struck by how much fighting is restrained, as if it followed unspoken gentlemanly rules (and it is mostly men). Stags lock horns, dogs bark and growl, birds flap their wings and gorillas beat their chests, but it's mostly bluff and bluster, stopping short of anything lethal. This makes sense, since fighting is a costly business, even for the winner. For Lorenz, the genius of natural selection is that it produced fixed patterns of behaviour or ritualised threats that enable animals to release and sense aggression without coming to harm.[41]

Many greeting rituals have a similar role, often straddling a fine line between friendly and threatening behaviour. The male cichlid flicks its tail, while the greylag goose performs its triumph display. We also see it in our own greetings, particularly between young men. The Maoris were known to greet unfamiliar visitors by throwing a spear in their direction, before sitting them down and rubbing noses. It also helps to explain the face-slapping of the Eskimos, as well as the likes of fist- and chest-bumping. One of my best friends and I usually punch each other on the arm. They're what Morris calls 'mock attacks'.[42] We tend to think of them as ironic, performed with a big grin; however, they might just have a more serious function.

Across the animal world, greetings also play an important part in ratifying social structures and status. Animals that live in groups tend to organise themselves around some form of hierarchy or pecking order. In a flock of chickens, which is where the term comes from, members will literally peck at each other to establish who's boss – who gets priority when it comes to eating, roosting or mating. The rest will eventually find their own position and the order will stabilise, removing the need for constant battling. It's the same with chimpanzees and the same with humans. Whether in school, work, politics or sports, our lives are full of hierarchy. We also have informal hierarchies among our friends and families based on who's the oldest, toughest, smartest or funniest. These pecking orders are signified by everything from our job title and uniform to our place at the dinner table or where we sit in the car (being the youngest of four, I was always in the boot).

Among chimpanzees, the alpha is greeted with a submissive pant and lowered head by other males, while the females present their rumps and allow him to mount them.[43] Among wolves, junior ranks will turn their heads and present their necks to dominant members of the pack. Dogs do something similar when they roll over at their master's feet. The basic point of all these gestures is the same: submission. By adopting a vulnerable position, the greeters are effectively surrendering themselves. As for our own greetings, we see the same with our rules of precedence and in the physical act itself, most obviously with the bow. The point is to make ourselves physically smaller, demonstrating our submission.

Conversely, if our status is equal, we express this through symmetry: whether shaking hands, hugging, bowing or kissing, the key is to match your partner. And, if the relationship isn't clear, we also use our greetings to assert our status, perhaps squeezing tighter or holding eye contact for longer. We see it with world leaders jockeying for position to make sure that their hand appears on top, or with Trump's 'yankshake'.

So where does the genital grab fit into all of this? Beyond the various rumours circling the internet, the fullest description of the Walbiri greeting comes from the Australian anthropologist M. J. Meggitt, who in the early 1950s spent fifteen months living with the tribe in the outback, producing the first full study of this little-known aboriginal group. Among his detailed descriptions of their daily life and customs, Meggitt recounts how, when men visited other groups, they would offer their penises, firmly drawing them down their hosts' hands. To refuse a penis was a sign of hostility, which could end in a deadly fight.[44] Other than underlining the importance of the ritual, though, Meggitt doesn't give an explanation of how it came about. It just seems like a very particular – and peculiar – cultural practice. Others have speculated that it has some sexual or homoerotic significance, perhaps revealing our bonobo side. But, from Meggitt's description, far from giving pleasure, it seems that penis-holding was an extremely serious business.

As it turns out, it's not bonobos that help us to understand this strange custom, but baboons. In certain respects, these Old World monkeys, with whom we share a common ancestor from around 30 million years ago, are more similar to us than chimpanzees, roaming the open savannah and living in large communities of around 100. They are also highly competitive and aggressive, with long canine teeth, and are best known for raiding picnics. There is, however, a notable exception to all this male–male friction and that's when they greet each other. On reunion, they will present and hold each other's genitals – what's known in the scientific community as 'diddling'. This curious behaviour got many scientists interested, and in the summer of 1983 a team of researchers spent four months studying a troop of olive baboons in the Great Rift valley in Kenya to try to make sense of it. In 2001, another team, from the University of Chicago, spent six months investigating the same thing at their local zoo.[45] What, they all wondered, was going on? Why would a baboon place his reproductive future literally in the palm of his opponent's hand?

Across the world, many animals (including humans) have traits or do things that make no evolutionary sense. The most iconic example is the peacock's tail. Consisting of 200 feathers and measuring up to 5 feet long, Darwin confided that the sight of them made him sick.[46] How could natural selection have produced something so wasteful and impractical? They clearly took lots of nutrients to grow and made it harder for the peacocks to get away from predators. For a long time, the best answer related back to Darwin's own theory of sexual selection, which suggested that males developed certain characteristics, such as long tails and colourful plumage, to attract mates. The size of the peacock's tail, like the bright colours, was just an example of competition getting out of control – like some of the dancing you see in a nightclub. But the problem was that no one could explain why females liked long tails or bright colours in the first place. Then, in the 1970s, the Israeli evolutionary biologist Amotz Zahavi came up with a compelling theory, which he dubbed the 'handicap principle'.[47] The point of the peacock's tail, Zahavi argued, was in its very dysfunction. 'Look at me,' peacocks are saying, 'I'm so strong I can survive despite this stupid thing.' The theory can be used to explain everything from long tails to giant antlers. We can see the same in our own species, with men buying gold watches, fast cars and designer shirts. 'Look at me,' they're saying, like a peacock, 'I've got so much money, I can waste it.'

Another famous example is the stotting of deer and gazelles.

Watch a film about the Serengeti and, at some point, you'll likely see a hunting sequence involving a lion and a gazelle. As the lion approaches its prey, keeping low in the grass, you'll no doubt be willing the gazelle to run for it. But, as it spots its killer, it does a strange thing: rather than sprinting for its life, it jumps up and down on the spot, as if goading the lion. Through this seemingly suicidal behaviour, the gazelle is essentially showing that it's so fit and fast that it can even give the lion a head start. The lion recognises this and so saves them both the hassle.[48]

As for the diddling of male baboons, the teams of scientists concluded that the ritual was a means of testing their relationships. By allowing each other to hold their most precious assets, they were showing complete trust and, equally, by not ripping them off, proving that they could be trusted – a way of initiating cooperation in a competitive world.[49] Put differently, it helps to overcome what economists call 'the commitment problem': how to know that a business partner or customer isn't going to pull out suddenly. It's partly why investors demand such high equity and estate agents require house deposits. Lacking such mechanisms, baboons put their genetic future on the line.

Zahavi's theory also helps to explain a number of other strange greeting rituals. There are capuchin monkeys, for example, which on meeting will often stick their clawed fingers up each other's noses and into each other's eye sockets. The deeper and longer they go, the closer they are, with some eye-poking sessions lasting for over an hour.[50] During my meeting with Bearder, he described one of the most remarkable rituals of all, which he'd observed in spotted hyenas. Also living in fission–fusion groups, on reuniting the hyenas stand parallel, top to toe, and display their erect penises, allowing each other to examine and sniff them. The really unusual bit is that the females do something similar. 'They have an enlarged clitoris, a sort of pseudo-penis,' Bearder explained, raising eyebrows on the table next to ours. 'Like the males, on meeting they also make them erect and present them to each other,' he continued. As with diddling and eye-poking, scientists have explained this risky custom in the same way: a ritual that quickly updates bonds and re-establishes trust.[51]

So, rather than being some weird cultural practice, the Walbiri greeting is just one of many that reveals our connections to the rest of the animal world. In fact, it seems likely that the custom used to be widespread. In ancient Rome, for example, soldiers would hold each other's testicles when pledging allegiance. There is also a clue in the word 'testicle', which shares its Latin root with 'testify': testis (one who bears witness). As the Dutch anthropologist Henk Driessen discovered through his study of male gesturing in rural Andalusia in Spain, it's not uncommon for men to grab each other's testicles as a friendly prank.[52]

All of which also helps to explain what was going through my mind when I decided to dunk my friend's head underwater at his swimming party, as described in Chapter 2. Just like the baboons and capuchin monkeys, the longer I held him down, the more I was proving that I could be trusted with his life (the fact that I'd got the wrong person is beside the point). There was also the time recently when, sitting outside our local pub, I noticed how one of the regulars greeted someone with the middle finger as he drove past. Rather than looking angry, the driver laughed and returned the gesture. Thinking back to the handicap principle, it made sense: our abuse can be a form of bond-testing. Just think how we often save our cruellest jokes for the people we're closest to or how much flirting involves teasing. Or there's the flow chart that deciphers how to tell if a British person actually likes you:

'Do they mock you, verbally abuse you, and make jokes at your expense? → Yes → You're their best friend.'[53] The British might be an extreme case, but the underlying principle is universal.

Each time we shake hands, hug or bow, we are putting ourselves in a vulnerable position, testing our relationships. It could be that the kiss is the riskiest of all, with de Waal characterising it as a sort of controlled bite.[54] No doubt we are a cooperative species, but competition is never far away, so we're always looking for signs of whether we can trust each other. It's why we pay so much attention to our body language: it's harder to fake. And, it's why, when it comes to greetings, we're all genital-grabbers of some sort.

7

It's All in the Mind

Juan Mann's life changed with a hug. Aged twenty-two and studying anthropology in London, he should have been in his prime. But things had begun to fall apart. His parents had divorced and his fiancée had broken off their engagement. He dropped out of his course, headed back home to Australia and ended up living as a recluse in a remote corner of town. Then, just as he'd hit rock-bottom, an old friend got in touch and dragged him to a party. And that's when it happened: a stranger came up and gave him a hug. His world turned. 'I felt like a king. It was the greatest thing that ever happened,' he recalled.

Having experienced the transforming power of the gesture, a few days later, on 30 June 2004, Mann made a sign – 'FREE HUGS' – and took to the streets of Sydney. Standing in a busy shopping mall, he initially felt terrified, lonelier than ever. But he resolved to give it an hour and eventually an old lady tapped him on the shoulder. She'd also been going through a tough time, having recently lost her daughter in a car accident and, just that day, her dog. But now that he was confronted with someone who actually wanted a hug, Mann suddenly felt uneasy about the whole idea. Nevertheless, he knelt down and put his arms out. It was one of the most awkward hugs he'd ever had, though the lady said that it had brightened her day and left smiling. It was enough to keep Mann going and soon he was giving out hugs every few minutes, each time feeling a surge of well-being himself.

Over the next weeks and months, Mann kept returning to the mall with his sign. He hugged anyone and everyone, from students and tourists who were after a photo opportunity to people suffering from deep mental trauma, apparently even bringing one man back from the brink of suicide. The good vibes were infectious. Even people who didn't stop for a hug were left smiling. It was 'fast-food emotion', Mann reflected. Soon others were joining him and, with the help of a music video, what became known as the 'Free Hugs Campaign' took off, spreading to more than seventy countries, including the likes of Norway, Jordan, Israel, Taiwan and even the UK (though it took a Portuguese waitress to get it going here).[1]

But while it was Mann and the music video that really got the Free Hugs Campaign going, none of this comes close to the efforts of a sixty-two-year-old Indian woman, Mata Amritanandamayi. Affectionately known as Amma (meaning mother), it's estimated that she has given out over 34 million hugs around the world, earning her unofficial canonisation as the 'Hugging Saint'. From a small fishing village in southern India, Amma was born into the 'untouchable' caste, so low that they were segregated from the rest of Indian society to avoid contamination. Growing up in a world of poverty, she was struck by the terrible suffering around her. As she went from door to door collecting scraps for her family's cows, she began to take food and clothes for the sick and needy, but also found herself embracing them and slowly became renowned for her healing powers.

Today, Amma runs a global charity, Embracing the World, which raises over $20 million each year in aid of various causes. But it's her hugs that continue to touch people the most. Regarded as a living god by many of her followers, she is said to have cured mental illness with her embraces. Thousands have flocked to her village to get a hug, while she herself travels the world giving them out in specially organised events. She's been known to sit all day and night, hugging over 20,000 people straight, only stopping for an insulin jab to treat her diabetes.

While Amma has millions of followers around the world, she also has her doubters, who dismiss her powers and all those who believe in them. 'Sad and pathetic! It's just a stupid gimmick,' says one.[2] In many ways, this summarises the other side of the reaction to the Free Hugs movement overall. Amma, Mann and their like are seen as a bunch of frauds peddling a load of hippy nonsense. Yet, although sceptical, I was keen to see for myself.

It turned out that I was in luck, with Amma scheduled to make her annual visit to the UK. All I had to do was turn up at Alexandra Palace in north London and queue for a token to ensure that I got a hug. I have to admit that, on the day itself, I'd started to doubt the whole idea: trekking down to London to join a queue so I could hug someone I'd never met. But as I was greeted by a smiling woman dressed in a white smock and orange sash – one of Amma's army of volunteers – my mood lifted. I was directed to the long line of Amma followers, all waiting for a token, and got chatting to a Bulgarian girl who'd heard about Amma from some friends after they'd attended one of her sessions. 'It was amazing,' they'd said.

A guy carrying a guitar overheard us. 'So you're Amma virgins,' he said. It turned out that he was a veteran, having travelled the world to see her. 'It seems like it's too good to be true, but it isn't,' he said. 'There really is something superhuman.'

Once I'd finally got my token and was guaranteed a hug, I was directed into the main hall, where I was assigned a seat – one of over 2,000. Amma was taking a break between the day and evening sessions, so there was time to get some food and explore.

Her followers looked like a mixed breed, coming in all shapes, colours and ages. There were a fair few straight out of *The Beach*, with their shaggy beards and flowing linen, a woman topped with a crown of tinsel and artificial flowers, but also people in national dress, business suits and everything in between. I got chatting to one of the volunteers, Mike, who, dressed in smart trousers and a tweedy jacket, didn't fit the mould. But it turned out that he'd been a follower for over thirty years, having converted to Hinduism and originally travelled to see her on the advice of a friend. He'd been suffering from terrible insomnia.

'When she hugged me, it was such a surprise that my whole body flinched,' he recalled, 'but then I just felt this incredible infantile love. I slept like a baby for weeks.' He ended up staying in her village for four years, until the heat finally drove him out.

But, as Mike explained, while Amma is considered a divine spirit – a great mahatma – she herself makes no claims about having any special powers. 'She just embodies the qualities we associate with motherhood of selflessness and unconditional love,' he said. 'It's an unbroken stream of humanity. She doesn't even stop for the loo. That alone is a miracle.'

Then there were James and Saumya, a couple of students who'd managed to sneak in at the end of the last session and had already had their hug.

'So what was it like?' I asked

Saumya's face lit up. 'It was just so happy, she's so blissful,' she replied.

James was less impressed. 'It just felt rushed. But I guess that reflects where I am right now,' he said, stroking his wispy beard.

But it turned out that James had cause to be more critical, since he was something of a hugging guru himself. It was his preferred method of greeting people, even strangers. In fact, as a sociology student in Manchester, he'd taken hugging to the extreme, organising an experiment with some friends – fellow members of the 'Zeitgeist Movement'. They stood in the street blindfolded with a sign at their feet: 'I trust you. Do you trust me? Hug?' I asked what the hugs were like, whether there were any memorable ones.

'Every one was different, but they were all good,' he said.

I asked if there were any awkward ones.

'Well there was this one bloke who hugged me and said, "I could stab you with a knife," which was a bit weird.'

The volume in the hall suddenly increased and I turned around to see that Amma had finally arrived. Dressed in a white sari, she sat cross-legged in the middle of the stage with a huge sign behind her: 'An Ocean of Love'. I took my seat as various people introduced her. We were reminded about Amma's charitable work, before she addressed us in her native Malayalam, translated by a man who sounded like God. We were warned about the travails of modern life, about how we are too focused on our material possessions, always looking at our gadgets, getting fat and lazy and not taking proper care of the world. We were told to love, encourage and respect each other, to eat more vegetables and learn the art of composting.

With her address over, Amma led a singing session, getting the audience to wave and chant. Now it was like Glastonbury. The rhythm of the drums and harmonium was relentless. Some people took to the floor, dancing and spinning like whirling dervishes. I still felt like an observer, finding myself thinking back to the sociologists' notions of 'interaction rituals' and 'collective effervescence'. Hindu, Christian, Muslim – it was all the same. The mood was then dialled back when we were invited by one of Amma's senior disciples to join him in meditation. It was new to me, but I tried to follow his instructions, observing my breath, imagining 'a sky full of pure white petals of peace' and focusing on the vibrations as he made a series of lawnmower noises. But I was finding it hard to let go. Having been sitting for over two hours, my back was playing up again, and I was feeling dehydrated and had a headache. I only hoped that one of Amma's hugs would sort me out.

Finally, just before 11 p.m., having been in the hall for seven hours, the stage was prepared for Amma to start hugging. Suddenly, it all got very efficient, as a BA-style announcement came over the tannoy, telling us that we'd be called up according to our tokens. I'd arrived early, so assumed that I'd be one of the first. But 'B5' flashed up on the screen – my ticket said S1. My heart sank. I'd heard that these sessions could go on all night. I showed my ticket to one of the volunteers who gave me a sympathetic look, saying I'd be waiting a while. 'But I don't understand, I got here really early,' I said, 'and I've got a long drive home.' But she said that there was nothing she could do: 'It's not in my gift. I don't know how Amma decides. It's a complex system.' It made it sound like there was some higher logic, which made me even more annoyed. But it didn't feel like the time or place to kick up a fuss. I wondered about asking her for a hug instead. She had the right gear on. But I told myself to calm down and enter into the spirit.

I sat back down and watched the big screen as Amma gave her hugs. She was surrounded by a scrum of helpers who called people up one by one. There's no doubt that everyone got a proper hug. Each one lasted over five seconds (I was counting – and by my calculations, multiplied by 34 million, that made over 2,000 days, or five and a half years of straight hugging). With every person, Amma smiled and pulled them in tightly, nuzzling her head against theirs, screwing her face into an expression of intense pain. It really did look like the sort of hug you'd get from your mum when your heart's just been broken, telling you that everything will be OK.

After an hour or so – somewhere around H2 – the screen with the numbers packed up and I found myself slipping to the back of the queue that was being called up to the stage, doing my best to avoid eye contact. My heart was pounding, as I felt a combination of guilt and fear that I'd be caught. But I made it to the stage and was called up to take one of the seats on the edge. I felt relieved but still guilty, like I'd somehow corrupted the whole thing. No one likes a queue-jumper. But I told myself to stop worrying and take in the moment. It was a different atmosphere on the stage, a cross between a royal court and a nativity play, with everyone huddled around the living god. There was one last heart-stopping moment when I was asked for my ticket and worried that I'd get hauled off the stage, right before Amma's very eyes. But it was just added to the pile.

I was instructed to kneel down and take off my glasses. The world became a blur. I could just about make out the man in front getting his hug. He went in for a second, surely pushing his luck. And then my time came. I shuffled forward until I was level with Amma and put my arms out. The world went quiet as she pulled me close, like I'd jumped into a pool. My head was right in her bosom. She had a unique smell, a combination of incense and baby lotion. She whispered into my ear, which I found a bit disconcerting, worrying that I was being cursed for pushing in. After my five seconds, Amma released. I looked up, like a child: 'Thanks very much,' I said. I felt something being pushed into my palm and was pulled away. I put my glasses on, left the stage, and looked at what was in my hand: a rose petal and a boiled sweet.

It might seem obvious that a hug should bring an extra sense of connection and well-being. But how do these feelings of warmth and euphoria come about? And how is it that the same gesture can easily backfire, causing an equal and opposite reaction? Much comes down to culture and the fact that our sensory experiences are mediated by the social context. Even in the United States, the head of a high school in New Jersey was forced to ban hugging as a greeting when some students complained that it made them uncomfortable. And I can't help thinking that if Amma had been a big man sitting alone in a room, my hug would have felt different. It's the same with other greeting rituals, and even with subtle variations in the strength of a handshake or the length of eye contact, which can provoke very different reactions.

For many people, from Plato to Descartes, Hindus to Christians, a core belief has been that our minds and bodies are somehow separate – the notion of the 'immaterial soul', that it is beyond our physical form. Our lives don't make sense otherwise. But gradually, neuroscience has challenged all this, showing that not only are our brains responsible for all our thoughts, feelings and experiences, but that they're highly structured, with different parts performing different functions, even to the extent that the idea of a single self has been called into question.

There's much about the brain that is still a mystery, but what we do know is that this walnut-shaped, wobbly mass is a chemical and electrical powerhouse, containing around 100 billion nerve cells or neurons, each with 10,000 connections – 'a disco of millions of dancing stars', as Amma's chief disciple had put it. Locked away in our skulls, with no direct access to the outside world, it's the control centre of our nervous system, processing all the information that comes in from our sensory organs and converting it into thoughts, feelings, decisions and actions. Pumped by various hormones and other chemical messengers, it gives us our emotions, dreams and imaginations, our fears and hang-ups, our consciousness, memories and personalities. It gives us the power of rational thought but also the capacity to believe in nonsense.

Understanding precisely which bit of the brain does what and how they relate to each other has been the main enterprise of neuroscientists. In recent years, the biggest development has come with the invention of 'functional magnetic resonance imaging' (fMRI), which shows brain activity by detecting changes in blood flow and lighting them up on a computer screen. It provides extremely detailed images and has enabled scientists to carry out ever more sophisticated investigations.

A leading figure in the field has been the neuropsychologist Professor Chris Frith, who helped to set up the Wellcome Trust Centre for Neuroimaging at University College London. In particular, Frith is renowned for his pioneering work on what's been dubbed the 'social brain', identifying the bits of matter that are responsible for how we think about and interact with each other.[3] I got in touch with Frith and was invited to meet him at the British Academy, the UK's national body for the study of the humanities. As we took the lift up to the fellows' room, he explained that after our meeting he was heading to a talk across the road at the Royal Society, the UK's academy of sciences. Through his work on the brain, he's one of the few people to have been elected as a fellow of both institutions. It was as if he still couldn't believe his luck, as he opened the fellows' room and did an introductory sweep. 'It's just sitting here empty,' he said. He offered me a coffee, but we were both defeated by the shiny new drinks machine, so turned to talk about something that made much more sense to him: the workings of the brain.

Frith had originally trained as a clinical psychologist back in the 1970s, but never practised, instead devoting his career to understanding how the mind and brain are related. To this end, he became fascinated with schizophrenia, which causes an altered mental state, and set out to investigate its biological basis. Using early brain-scanning techniques, Frith showed how schizophrenia was a brain condition, partly caused by a disconnection between different areas and an abnormal blood flow. 'I like to think that I helped to change the way that society views schizophrenia,' he said. By a happy twist of fate, at the time Frith was studying schizophrenia he met his long-time collaborator and wife, Uta Frith, who happens to be one of the leading experts on autism. In parallel with Chris Frith's work on schizophrenia, she transformed our understanding of autism, showing how it is not the result of cold parenting ('refrigerator mothers'), as had been contended, but also a condition of the brain.

In particular, she highlighted how autistic people find it difficult to perceive and interpret what other people think and feel. They lack, or at least struggle with, what's known as a 'theory of mind': the ability to take on the mental state of another person. One of her students, Simon Baron-Cohen, who now runs the Autism Research Centre at the University of Cambridge, developed the 'Sally–Anne test' to assess people's theory of mind. Participants are introduced to two puppets (Sally and Anne) and then told a story which involves Sally taking a marble and putting it in her basket, before heading off for a walk, leaving her basket behind. With Sally gone, Anne takes the marble and puts it in her own box. When Sally returns, participants are asked: 'Where will Sally look for the marble?' Whereas Sally would naturally look in her own basket, Baron-Cohen found that around 80 per cent of autistic people answer that she will look in Anne's box – they fail to adopt Sally's mindset. It's this 'mindblindess' that can make autistic people seem so single-minded, but also struggle to relate to others.

As Chris Frith explained, having a theory of mind is integral to our social interactions. However flawed, in many ways, we're all instinctive mind-readers, constantly trying to work out what's going on inside each other's heads and respond accordingly. In our own self-centred way, though, a lot of the time we're attempting to gauge what other people's feelings and intentions towards us are. We might be intensely social creatures who rely on cooperation, but, as we know from the last chapter, we have our own interests at heart (even if they're manifested altruistically), which can easily be turned against each other. From the moment we meet someone, we're alert to the external signs of what's going on in their minds, their facial expressions and body language, and asking questions about what they think and feel. While we can all struggle, these are things that the autistic brain fails to process in the same way. As Uta Frith concludes, it's as if autistic people are missing the 'tiny gadget' that enables the rest of us to naturally take into account what someone else wishes, thinks and believes.[4]

Drawing on his wife's research, Frith showed how schizophrenia is also partly characterised by an impaired theory of mind. As well as losing touch with their own thoughts and feelings, sufferers struggle to monitor the intentions and beliefs of others. Gradually, though, with the development of more sophisticated fMRI techniques, Frith began to look at social behaviour more generally. He found that a particular part of the brain becomes active whenever we try to think about other people's mental states: the medial prefrontal cortex (mPFC). Sitting just behind the forehead, this is part of the outer rind of grey matter, known as the cortex, which in turn accounts for most of our recent brain growth and gives us our highest mental powers.

At the same time that Frith was mapping the social brain, an Italian neuroscientist based at the University of Parma, Giacomo Rizzolatti, made a discovery that's been heralded as one of the biggest breakthroughs in modern science: 'mirror neurons'. As the name suggests, these are neurons that fire both when we do something and when we observe the same action performed by someone else. As Frith explained, it all came about by accident, 'what we like to call serendipity'. Working with macaque monkeys that had electrodes fitted to their brains, Rizzolatti and his team were investigating how certain neurons relate to different body movements, such as grabbing and eating a peanut. To their surprise, they noticed how some neurons fired not only when the monkeys picked up the peanuts, but when the experimenters picked them up as well. As Frith remarked, in many ways it was common sense: it stands to reason that the brain is active when observing others. But what Rizzolatti's chance discovery revealed is that there is an underlying neural mechanism: special neurons that both 'see' and 'do'.

While their precise function is still debated and specific mirror neurons haven't yet been found in humans (to do so would require directly wiring up the brain), fMRI studies have revealed areas of probable mirror-neuron activity, particularly in the frontal and parietal cortex, suggesting that we do indeed have a general mirror system. Moreover, these are not just parts of the brain associated with body movement: they are also involved in our emotional states. It would certainly help to explain why we wince when we see someone walk into a door, cry when others cry or jump for joy when our favourite player scores a goal: we literally feel some of their pain or pleasure. It also perhaps explains what brought a smile to the people who saw Mann giving out his free hugs. Whether we like it or not, we find ourselves taking the mental perspective of others. More generally, the discovery of these 'Gandhi neurons', as they're also known, has blurred the boundaries between ourselves and others, suggesting that our brains – our minds – are interlinked.[5] As the Bulgarian girl put it, 'science is proving a lot of the spiritual stuff' (or at least making more sense of it).

This urge to imitate serves an important social function. When analysing political summits or scenes from *Big Brother*, pundits often draw attention to whether the participants' body language is in sync. When we're talking to someone, particularly if it's going well, we tend to match our postures, expressions and even our accents. It's what's known as the 'chameleon effect' and helps to explain why long-term couples can end up with similar wrinkles. Some have contended that this is all part of our effort to work our what's going on inside other people's heads (the theory of mind again), but, as Frith explained, a more likely explanation is that we're simply building a sense of rapport.

Most of the time mimicry happens automatically, but when it doesn't and our body language diverges the results stand out. To illustrate the point, Frith demonstrated how, when he doesn't agree with someone in a meeting, he finds himself sitting on his hands. Like many, I end up crossing my arms or gripping the chair. Body-language experts interpret these as barrier signals or displacement actions, but we may partly be suppressing our urge to imitate. It can be the same with greetings. If someone sticks their hand out it's almost impossible not to reciprocate. And, if we don't, it's a big deal. Clearly, this has a lot to do with custom and politeness, as well as force of habit, but the fact that it's so involuntary suggests that something else might be going on.

In recent years, though, Frith's own attention has increasingly turned to the nature of consciousness.[6] What's the point of being aware of our experiences and being able to reflect on it all? For Frith, the main reason we have consciousness is precisely that we're such social creatures. By being conscious of our actions, we're able to explain and justify them, which enables us to build understanding and consensus among our peers. So, when we get in a tangle over our greetings, say, we can at least apologise or make a joke of it. While all this might be true, it can often feel as though the bigger point is not so much that our consciousness enables us to explain our actions after the fact, but that it leads us to question them beforehand. It's surely a defining element of the human condition: we don't just make decisions but wonder if they're right. It's what's known as 'metacognition' – thinking about thinking – and can get us into a real mess, sending us into a spiral of indecision and anxiety.[7] Again, it seems that the mPFC is largely to blame.

Compounding all this is the fact that our social brains don't just question us but second-guess others too. We look for signs of what is going on inside other people's minds so that we can pre-dict what they're going to do, particularly in relation to ourselves. When we get it wrong or make a 'prediction error', our brains are updated. It's another crucial part of the learning process, and neuroimaging shows that the mPFC has a central role. Through our experiences, we're building a store of reliable signals, a sort of field guide to human behaviour, which we're constantly editing.

In many situations, our own decisions are based on what we think others are going to do, so we can either cooperate or outwit them. The problem is that we know that we're all at it, which complicates matters. As Frith explained, bringing up a cartoon of two children on his computer, 'It's like a game of hide-and-seek.' You might choose to hide behind your favourite tree. But then you remember that your friend knows that you always hide there, so you choose the garage instead. But, thinking about it further, you realise that your friend will know that you know that they know your favourite spot is the tree, so will probably look in the garage – so maybe the tree is the best spot after all. And then you realise that they know that you know that they know that you know... And so on. This is what's known as 'recursive thinking' and our interactions are full of it. As Frith explained, most people are capable of one or two levels of recursion before it gets too much.

'Sorry, we seem to be getting off-topic,' he said.

'Not at all,' I said, thinking that we were actually getting to the nub of the matter. 'It's just the same with greetings – you know, "one kiss or two?" and all that.'

I wondered if all this talk of recursive thinking could help us out with our greetings, mentioning how the best that the body-language advisors could come up with was to follow the other person's lead.

'It does seem like the obvious solution,' he said, before thinking for a moment. He then explained how in situations where we're looking to cooperate or synchronise our behaviour, the less recursion the better. We want to do what the other person wants us to do and stick with it. 'I suppose the first thing to ask is whether the person is in my group or not,' he concluded. 'Well, it works if you know the etiquette,' he smiled.

When it comes to those all-important first impressions, there's a lot going on beyond getting our greetings right. From the experiments carried out by Alexander Todorov and his team at Princeton (discussed in Chapter 4), we know that we make judgements about people within a split second, particularly in relation to how trustworthy they are. Living in a world where our interests can clash, this makes sense, or at least evolution seems to have primed us this way. So is there a part of the brain – a neural correlate – that is responsible? To find out, Todorov got volunteers to look at his range of faces (from trustworthy to untrustworthy), while lying inside an fMRI scanner. As expected, he found that a particular bit of the brain lights up when people are confronted with untrustworthy faces: the amygdala (pronounced 'a-mig-der-ler').[8] This is an almond-shaped structure that is divided between the two hemispheres of the brain and not much longer than a centimetre. It is located down in the limbic system, which is an older region of the brain, most commonly associated with our automatic reactions. While the amygdala has several functions, it's most commonly associated with generating our emotions. For example, it's important in fear, responding to external signs of danger and sending projections to other parts of the brain, including the mPFC. It's the brain's internal alarm, spurring us to act. So it makes sense that the amygdala fires when we're confronted with an untrustworthy face.

The amygdala is also involved in perceiving and processing people's expressions and body language, particularly when they seem threatening. More generally, it responds to anything that's unfamiliar or unexpected, constantly keeping us on guard (remember Bertrand Russell's comment about how we tend to fear and hate what is unfamiliar). This helps to explain why we tend to find average-looking faces trustworthy: they are an amalgam of what we know.

More fundamentally – and controversially – the workings of the amygdala help to explain racism and prejudice as a whole. Whether we like it or not, it seems that our brains, kick-started by the amygdala, are constantly looking for mental short cuts, processing information and making instant judgements, particularly when it comes to assessing whether someone presents a threat. Maybe this is a legacy of our days on the savannah, when danger was never far away, or perhaps it's just a function of the fact that we're forever finding ourselves in new situations. Whatever the reason, we tend to categorise and package people, automatically forming first impressions, gauging whether they fit the norm and, thereby, how trustworthy they are. As Frith explained, it's what academics call 'prior expectation' – otherwise known as stereotyping or prejudice – and it's a big part of how the brain works. We're looking for who's in and who's outside our group. Racism might be the crudest form, but it seems that we are all prone, not just the people who act on it.

Studies have shown that, from the age of three months, infants already have a preference (measured by their attention span) for faces from their own racial background.[9] And later in life we remain alert to racial differences and, consciously or not, use them as mental short cuts. Stepping into a potential minefield, a team of social psychologists led by Allen Hart at Amherst College in Massachusetts investigated what happens in the brain when we see someone from a different race. Volunteers were recruited to look at images of white and black people while inside an fMRI scanner. As Hart predicted, there was an increase in activity in the amygdala when participants were shown pictures of people from a different race to their own, despite the fact that none of them admitted to feeling any kind of racial prejudice.[10]

So what should we make of all this? We should first note that, like all bits of the brain, the amygdala does not operate alone, but sends signals that are mediated by other areas, including the cortex, which have higher mental functions. In other words, we can – and do – refine and override the amygdala's initial response. Second, the amygdala works from memory and experience, so is conditioned by social norms and attitudes, both good and bad. Moreover, experiments have also shown that the amygdala does not fire in the same way when people are shown photos of people from different races who are familiar, such as film stars. In short, familiarity breeds trust. Finally, and perhaps most importantly, like all these things, just because something appears to be natural or hard-wired does not mean it is right. It's simply another case of 'know thyself' and being aware of the fact that our brains are prone to calling up stereotypes and making snap judgements. We also know that we have the brainpower (our cortex) to over-rule them. The issue is that this can take extra effort, as we feel ourselves doing battle, reconciling different mental processes: our reflex judgements and critical thought, or what Daniel Kahneman calls 'fast' and 'slow thinking'. So when we feel conflicted or find ourselves fighting our own thoughts, we know that this is a neural process, with different parts of our brain churning away. While our amygdala helps to keep us out of harm's way, it's the extra mental work, when we engage our consciousness and critical thought, that we might call progress or civilisation.

The flip side of all this is our desire for social contact and acceptance. Our brains may be constantly on guard against outsiders, alert to abnormalities, but as the Princeton social psychologist Susan Fiske says, the first thing that drives our social behaviour is belonging: to feel part of a group, whether it's our family, friendship circle, tribe, sports team or political party.[11] We might all need our own space at times or want to hide from humanity altogether, but ultimately most of us crave human contact. And when we're starved of it, our brains suffer. We feel lonely and depressed. It's why solitary confinement is considered one of the worst punishments and banned in many countries. Conversely, when we are with our family and friends, we feel a sense of well-being (at least most of the time). For many, it's what makes us get up and go to work each day. It's not the job we care about but the people around us. So it's hardly surprising that our biology gives us a helping push, generating the sensations that motivate us to come together and avoid isolation.

Again, much of it comes down to a particular region of the brain: the basal ganglia or, more specifically, its biggest part, the striatum. Situated at the base of the forebrain, down in the limbic system, the striatum sits at the centre of what is known as the brain's reward system, mediating all those good feelings and making us want more. As well as being linked to other parts of the brain, including the mPFC, it's full of neurons that are receptive to an essential ingredient of the reward system: dopamine. Popularly known as the happiness drug, dopamine is a chemical messenger (a neurotransmitter) that sends signals to other neurons, particularly in the striatum and mPFC, and gives us feelings of pleasure. It's the massive dopamine rush that makes taking heroin feel so good – and makes users want more.

The reward system is also activated when something good happens, such as winning the lottery, being told we've got our dream job, finishing the washing up or getting a like on Facebook, motivating us to do the same again. And in just the same way, it's the striatum and dopamine system that fires when we have a positive social encounter, from a friendly smile or hello to something more intimate. Crucially, though, these neural rewards are mediated by expectation and, in turn, social norms. So when our interactions go to plan, we feel content. But if we experience a nice surprise, maybe a smile from an attractive stranger, bumping into an old friend or praise from our boss, our striatum and dopamine indicators spike. It helps to explain why Mann felt so euphoric when he was first hugged at that party: his expectations had been rock-bottom. Conversely, when our hopes get knocked, maybe a date goes badly or our Facebook comment doesn't get any likes, our dopamine levels drop and we feel flat.

On the other side, if we realise that we've made a social faux pas – perhaps forgetting someone's name or going for a hug only to sense the other person squirm (or whatever greetings blooper you can think of) – we feel embarrassed. Worse, we start to blush, which makes us even more self-conscious and blush more. And, in the extreme, we end up becoming flustered and lose all mental faculties, descending into a blundering mess. All this can seem pretty cruel for a social slip-up. Moreover, it seems that Darwin was right when he suggested that blushing is the 'most peculiar and most human of all expressions'.[12] Other animals may experience the physical symptoms (a sudden rush of blood to the face), but it's likely that it's only us in whom it's triggered by embarrassment.

The precondition for all this is the ability to imagine what other people think of us; without it, we wouldn't be able to care. Generally, we blush when we feel ourselves being evaluated by others – just when we don't want the extra attention. But while blushing is uncomfortable, it can actually serve an important function. Whether we like it or not, it advertises the fact that we're aware we've made a social misstep. Even though we've breached the rules, we at least show that we didn't mean to. It's a silent apology: a flashing sign saying, 'I know I've made an idiot of myself but you can still trust me.' And while it might feel unpleasant, it seems that when we've made a blunder we're better off blushing than not doing so, since studies show that people will view us less negatively.[13] So when you next find your face on fire and are reduced to a floundering mess, remember that it has an important social function. At least this should prevent you from trying to stop it, which only makes it worse.

In the extreme, our social transgressions can lead us to be rejected or excluded. Worse than embarrassment, we feel pain. We often say that we're 'hurt', comparing the feeling of rejection to being punched in the stomach or hit by a car. We might be charged with emotion, but it turns out that these analogies might not be so melodramatic. In another crafty experiment, a team of social psychologists led by Naomi Eisenberger at the University of California, Los Angeles, investigated what goes on in our brains when we're socially excluded. She had volunteers take part in a virtual game of catch while lying in an fMRI scanner. First, they watched a game on their screen between two other players, whom they believed were also taking part in the experiment, before being invited to join in. As usual with psychologists, all was not as it seemed. In reality, there were no other players, just a heartless computer that was programmed to stop throwing the ball to participants so that they felt left out. Even during this harmless little game, Eisenberger found that the patterns of brain activity were similar to those that occur when we experience physical pain.[14] When it comes to greetings, all this explains why a snub really can hurt.

Much of the above underlines the power of custom and

conformity – or, put another way, of etiquette. In short, the brain's reward (and penalty) system motivates us to keep within the boundaries of our social norms. This urge to fit in can exert a strong influence, even suppressing our better judgement. In the 1950s, Solomon Asch, whom we met in Chapter 4 with his research on social judgement, conducted a series of experiments investigating the effects of group pressure. Volunteers were invited to take part in a 'vision test' in which they, along with seven other participants, were shown a single line along with a group of three other lines and asked to choose which of the three lines matched the single line. The answer was obvious, but what the volunteers didn't know was that the other participants were part of the ploy and had been instructed to choose one of the other, mismatching lines. With the volunteers going last, Asch found that nearly a third of them followed the rest of the group, despite the fact that they were clearly wrong.[15] It's possible that this was merely a sign of more conformist times, but the test surely points to something that we've all experienced: the power of the group and our desire to fit in.

In the end, it seems that the brain's reward system has largely evolved to turn us into sheep. It's why organisations are beginning to realise that the dreaded group brainstorming and Post-it note sessions might not be such a great idea, if we're really after the most original and critical thinking. Faced with our peers, we tend to go with the flow. It's why members of even the most radical countercultural movements end up establishing and following a new set of patterns and customs: to recreate a sense of belonging. Mods, goths, punks, indie kids, hipsters – by definition, they're all conformists, really. And I now realise it's why, despite my own convictions, I ended up becoming a double- and even a triple-kisser.

———

When it comes to greetings, it's not just our brains that are social, but our bodies too. Starting from the beginning, we first identify the person we want to greet. Generally, we do this using our dominant sense: vision. As the saying goes, 'seeing is believing.' But while we tend to think that seeing is all about the eyes, it's actually the brain (mostly the occipital lobe at the back) that's doing the work, processing all the visual information that's coming in and forming an image. Different parts handle different elements, such as size, colour and movement. But it turns out that there's also an area that plays a critical role in facial recognition: the fusiform gyrus, which sits across the occipital and temporal lobes. If damaged, people can end up with a subtle but debilitating condition called prosopagnosia – more commonly known as 'face blindness'. While sufferers can otherwise see normally and identify a particular eye or nose, they struggle to put it all together. In the extreme, they can't recognise their friends or family, or even themselves, and end up blanking people, not recognising who's saying hello or reintroducing themselves to someone they've just met. It's likely that more than one in fifty of us may be affected to some degree.

All being well, though, if we've made a positive identification, we generally make eye contact – the initial offer. As another saying goes, eyes are the 'windows to the soul'. In fact, the most basic clue to what someone else is thinking is where they're looking. While other animals, such as chimpanzees, also follow each other's eyes, indicating a basic theory of mind, humans are particularly attuned, helped by the fact that, uniquely, our coloured irises are surrounded by a large area of white: the sclera. It's even suggested that the sclera evolved for this very reason, as a social signal for others to detect where we're looking.[16] When it comes to our greetings, it's obvious when someone has returned our offer, showing that they are now thinking about us.

Aside from demonstrating our undivided attention, when we look into each other's eyes we're also picking up physical information, not only about shape or colour, but about pupil size. The field of 'pupillometrics', pioneered in the late 1950s, has shown how we generally find faces with bigger pupils more attractive, friendly and trustworthy.[17] It helps to explain why the wicked witch always has beady eyes, while Snow White and Sleeping Beauty have supersized pupils. As more recent studies have shown, though, perhaps the key point is that our pupils dilate when we are emotionally aroused – in other words, when we are interested in something.[18] So, in our egotistical way, when we notice that someone has big pupils, we take it as a sign that they are interested in us and hence respond more favourably.*

A more obvious sign that someone is pleased to see us is the smile. While we tend to think of the smile as a universal expression of joy, there are significant cultural variations in how we use and interpret the gesture. In short, we don't need to feel happy to curl our lips. When we greet someone, for example, we can feel a whole range of emotions, from excitement and relief to dread, guilt and panic. Yet we invariably smile. So should we ever take people's expressions as a true sign of how they feel? It's a question that the American psychologist Paul Ekman set out to answer back in the 1960s. Starting out, he'd expected to confirm Ray Birdwhistell's view that all expressions are learned and depend on our culture (as discussed in Chapter 5), but in the end he found himself confirming what Darwin had concluded

---

* Though it's worth remembering that pupil size is also a function of light level.

150 years earlier: that our basic expressions are universal, corresponding with our core emotions (happiness, anger, disgust, sadness, fear and surprise). Using photos of each expression, Ekman found that people from across cultures matched them to the same emotion. To make sure, he even visited an isolated community in Papua New Guinea that had had no outside contact. Apart from some confusion over fear and surprise, he got the same results, demonstrating that our expressions really are a true sign of how we feel.[19]

So what about all the inconsistencies – all those phoney and forced smiles in family photos or when we bump into our boss on a night out? To square the circle, Ekman came up with the notion of 'display rules', capturing the fact that we manage our expressions according to certain cultural codes and expectations. To demonstrate, he asked a group of American and Japanese students watch a film of accidents and operations. Recordings showed that when they watched the film alone, the students made the same range of expressions. But when an observer was present, the Japanese participants invariably masked their negative emotions with smiles, while the Americans continued to show their horror.[20] As the psychologist David Matsumoto later concluded, the differences came down to the fact that Japanese culture is more collectivist, emphasising group harmony, so people tend to suppress expressions that might upset the group.[21]

Yet, even though we can easily smile, we can't so easily fake a real one. In 1868, the French neurologist Duchenne de Boulogne was investigating how our facial muscles change our appearance by stimulating them with electrodes. When he activated the muscles around his subject's cheekbone, forcing a smile, he noted that he didn't really look happy. So he told him a joke, which revealed a clear difference: the muscle around the eyes (the orbicularis oculi) also contracted. As Duchenne concluded: 'Its inertia in smiling unmasks a false friend.'[22] It led Ekman to christen the smile of enjoyment the 'Duchenne smile'. As he observes, in a way that's sure to get people paranoid, when happily married couples meet at the end of the day, their smiles involve the muscle around the eye, but it is missing from smiles of unhappy couples.

If we need to, though, there are ways of activating the orbicularis oculi and simulating a Duchenne smile. The basic method is to squint and feel your cheeks rise and then part the lips, before reopening your eyes. Or, to make it easier, smile while biting a pencil. Amazingly, Ekman found that, by forcing a smile in this way, we activate the same parts of the brain as when we smile spontaneously.[23] In other words, the act of smiling can actually make us feel happier. Similarly, in a subsequent experiment in the 1980s, the German social psychologist Fritz Strack found that when he got participants to look at cartoons while biting a pen they found them funnier.[24] All of which led Ekman to propose that holding the Duchenne smile for twenty seconds could be prescribed as a form of therapy.*

As to the approach stage of greetings, this involves crossing our zones of personal space – those invisible bubbles we met earlier. Neuroscientists have since shown that we really do have a sense of interpersonal space and that it's modulated by the amygdala. While the distances might vary with culture, people with damaged amygdala can end up without any sense of personal space at all, with one sufferer walking all the way up to an experimenter to the point of touching without reporting any discomfort.[25]

---

* It should be noted, though, that a number of researchers have replicated Strack's experiment and failed to get the same results. In the end, smiling with a pencil in your mouth might just make you look stupid. All in all, scientists still have some way to go before they've fully cracked the relationship between our expressions and our emotions.

This brings us to the contact display, which engages what's arguably the most complex and powerful sense of all: touch. Our tactile sensitivity is immense. The smallest touch can trigger intense feelings of intimacy and excitement or make us recoil with horror. Imagine, for example, how the British tennis star Andy Murray felt when he was woken by someone stroking his arm. Had it been his wife, no doubt it would have been a happy start to his day, but it wasn't: it was his hotel maid, who subsequently stalked him across Europe. Clearly, as with all our senses, we process touch very differently according to the situation and context. As Frith explained, our 'bottom-up' sensory inputs are modulated by 'top-down' mental processes that, crucially, introduce past experience and expectation into the equation. Yet, when it comes to our greetings, for all the variations, it's striking how some form of touch is so common and integral, even with strangers. And, significantly, scientists have recently demonstrated what might seem obvious: how we moderate our touch according to the closeness of our relationships.[26] Our greetings might throw up some interesting anomalies but, as a rule, the closer our bonds, the more physical they are.

We all have a basic need for physical contact, which may explain why touch is the first sense to develop. By the time we've made it out of the womb, our touch organ – the skin – is by far the largest sensory organ, comprising some 18 per cent of our body mass. As you'd expect, the densest areas of touch receptors are in our hands, lips and tongues. Of course, all of this is vital for our survival, steering us away from harmful things, and allowing us to explore and manipulate the world around us. But, as experts in the Group for Research in Affective Somatosensation and Pain (GRASP) at Linköping University in Sweden conclude, the skin is fundamentally a 'social organ'.[27]

Accordingly, the brain processes human touch, especially from others, very differently to when we come into contact with an inanimate object. And try tickling yourself. Sadly, you can't – it's an exclusively social affair. As the neuroscientist Robert Provine found when he stepped out of the shower one day, the best you can get is if you tickle the underside of your left foot with your right hand, which almost tricks the brain into thinking that someone else is doing the tickling.[28]

From the moment we can sense the world, social touch plays a key role in our well-being. As we know, it's our mother's embrace that comforts us, making us feel safe in the big bright world. More generally, though, touch is vital to our emotional and physical development, with numerous studies showing that infants who are severely deprived of tactile stimulation, such as in understaffed orphanages, have a broad range of developmental problems, from impaired growth to slowed cognition. Of course, these may result from a more general lack of social stimuli, but it's been found that in most cases twenty to sixty minutes of gentle massage is enough to reverse the damage.[29]

Beyond childhood, touch continues to play an important role in maintaining our relationships, particularly with our partners. More generally, though, platonic touch can be a key element in our day-to-day interactions – and in ways we don't always realise. It turns out that when we say 'I'm touched' to express our appreciation, we might not be being entirely metaphorical. One of the earliest demonstrations was in 1976, when researchers at Purdue University in Indiana got the librarians to take part in a simple experiment. When returning library cards, they were instructed to place their hands over the palms of some students, thereby making physical contact, while avoiding touching others. Outside the library, a researcher approached the students and asked them how they rated the librarian. They found that those who had been touched gave more positive feedback, even though over half of them weren't aware of anything different.[30] Since then, a string of studies have shown how this kind of minimal touching can have a big impact. It's been found that waiters can increase their tips by up to 15 per cent and charity workers can expect bigger donations with just a short touch to the forearm.[31] In one experiment, participants were more likely to take financial risks if they received a light pat on the shoulder from the coordinator.[32]

Before we interpret these results too broadly, though, we should take into account a couple of things. First, most of the tests were carried out on American college students, so don't allow for much cultural variation. As a study of patterns of inter-personal touch by a team at the University of Oxford recently showed, people from the UK don't really like to be touched at all by strangers. And, second, as the same study demonstrated, all touch is not equal. Using body maps, the researchers produced a 'touchability index', which showed how, across cultures, there are certain areas that are out of bounds to all except the people we are closest to. As you'd expect, these include the genitals and buttocks, while the face, head and neck are also restricted areas. Nevertheless, even though we might not all have the healing powers of Amma, as the team noted, touch does have a special effect in our relationships.[33]

But what about ritual touch, touch that we expect and do as a matter of course, such as a handshake, for example? Recently, a team of social psychologists at the University of California, Berkeley, spent a lot of time watching basketball, carefully recording particular moves. They weren't looking at the players' shooting or passing skills, but at the number of times they deliberately touched their teammates: all the high fives, chest bumps, team huddles, and the like. They found that the players and teams that touched the most were the best performers over the season, leading to the conclusion that this kind of ritual touch really can enhance our relations.[34] Obviously, this could all be correlation (the best players score more and so touch more), but it would help to explain why body-language advisors, politicians and business leaders alike make such a big thing about the handshake.

What scientists do know is that touch can stimulate a range of neurotransmitters and hormones in different parts of the brain, which make us feel good. As we know, dopamine plays a big role, but there are others, including endorphins, which also give us drug-like highs. It's endorphins that kick in when we go running, masking the aches and making us feel like Rocky. Less painfully, they're also released during moments of physical intimacy and help to explain why primates spend so long grooming each other and cats love to be stroked.

When it comes to our social relations and feelings of attachment, though, another hormone is widely seen as playing the leading role: oxytocin. Produced mainly in the limbic system, when released oxytocin activates special receptor neurons in different parts of the brain, changing the way we feel and behave. Early tests on rats and sheep showed that an extra dose increased maternal care and strengthened the bonds between mates. Subsequent experiments with humans demonstrated a whole range of positive effects, from reducing social anxiety to boosting feelings of friendship and empathy. Most notably, scientists found that oxytocin can increase trust, showing how a squirt up the nose can make people perceive faces as being more trustworthy or inclined to make financial investments in an anonymous partner.[35] All of which has led Paul Zak, one of the leading researchers in the field, to characterise oxytocin as a 'trust switch' in our brains. And, as he explains, it also triggers a release of dopamine, which makes our trusting behaviour feel good. In this way oxytocin encourages our most positive social behaviours, generating feelings of generosity and compassion.[36]

Given all these apparent benefits, it's no surprise that public interest in oxytocin has been high, with clinicians investigating the possibilities of using it to treat a range of conditions, from autism to depression. Although doctors can't legally prescribe it yet, for around $30 you can get a bottle of OxyLuv nasal spray on Amazon. Most customers seem pleased with their purchase, reporting various benefits, from reduced anxiety to improved marital relations. In fact, one buyer was so impressed that he started to pour it into his coffee each morning – 'Just a bit, but what a difference.'[37] (It turns out that this might not be the best idea, though, since oxytocin apparently breaks down as soon as it hits the stomach, which is why it comes as a nasal spray.)

But you don't have to resort to buying the stuff to boost your oxytocin levels. Simple human touch can trigger a surge. In another experiment, Zak found that a fifteen-minute massage stimulated oxytocin and led to participants being more trusting in financial games.[38] And a hug can also do it. In fact, oxytocin has been nicknamed the 'hugging hormone', spurring on the Free Hug activists and leading psychologists, particularly in the United States, to laud the physiological benefits of hugging. So it turns out that the likes of Mann and Amma might be spreading the love after all – and surely getting an extra kick themselves.

Moreover, it doesn't take much of a leap to wonder if oxytocin helps to explain all those experiments demonstrating the power of touch. For his part, Zak suggests that we do indeed get a small charge in response to a range of social signals of trust, such as a handshake.[39]

Before we get carried away, though, patting and hugging with abandon, expecting a boost to our relationships and well-being, a few qualifications are in order. More recently, some scientists have been urging caution, suggesting that the effects of oxytocin are more complicated than the likes of Zak suggest, especially when it comes to humans.[40] In fact, when I met with one of Frith's colleagues, Uri Hertz, who has investigated how oxytocin affects group interactions, he explained that an extra dose can also give rise to more negative tendencies, such as in-group favouritism, even distrust of outsiders. Perhaps this accounts for the one-star reviews of OxyLuv on Amazon, with people complaining about a whole range of unexpected effects, from becoming overly emotional to feeling aggressive and irritable. Moreover, there is the great charge of reductionism – reducing complex behaviour to a single factor, such as a neat chemical reaction. Hertz was politely sceptical when I tested my armchair theory that touch can trigger oxytocin almost like hitting the button on a drinks machine. As he explained, in processing touch, our brains also take into account context and our own experiences and expectations, as well as other sensory inputs. Given the wrong circumstances or person, touch can easily make us feel uncomfortable, even distrustful.

That said, I still wonder. Could it be that those of us from 'low-contact' cultures are missing out, and that more tactile people and societies are somehow better off? Certainly, beyond the likes of Mann and Amma, there are a host of people – including scientists – who believe that we're in danger of depriving ourselves of our most basic sense. Maybe the Victorian etiquette advisors did us a disservice with their emphasis on self-control and restraint, leaving a legacy of tactile poverty. At home, we baulk at 'unnecessary physical contact' and 'public displays of affection'. I do myself, but I've also been struck by how other cultures seem less allergic and more connected. There was a particular moment when I was standing in a town in Eritrea as the students poured out of school. They greeted each other with handshakes and shoulder bumps, and walked arm in arm. Even the teenage lads were draped around each other as they sauntered along. One boy ran over and poked me to see if I was real, before kissing me on the hand and running off.

I'm wary of taking this too far. After all, dubbed the North Korea of Africa, Eritrea isn't exactly renowned for being the most open country in the world. With its repressive regime, many Eritreans just want to escape. But at an interpersonal level, like other cultures, they were visibly more connected than we are back home. In contrast, in the UK we can recoil at the slightest physical contact. No doubt we can blame the Victorians, but some commentators say that we've become even more touch-phobic in recent times. We're paranoid about a pat or hug being misinterpreted, even landing us in court. From our workplaces to our schools, we're warned to keep touching to a minimum. Of course, we need to protect against inappropriate touch and harassment, but some argue that we've gone too far and are starving ourselves of an essential element of interaction. You only have to look at our closest animal relatives, joined in mammoth grooming sessions or fixed at the lips, to see how important touch has been in forming our bonds. And as neuroscientists have shown, the effects are mental, stimulating various mechanisms in the brain. Personal and cultural hang-ups aside, we're biologically primed for social touch. All of which underlines the importance of our greeting rituals, which are often touch's last stand.

It turns out that hormones might also account for some of the individual differences in how we greet. In particular, a team of social psychologists led by Frank Bernieri, director of the Interpersonal Sensitivity Lab at Oregon State University, investigated the effects of testosterone on our social encounters. Produced mainly in the testicles (and, to lesser extent, in the adrenal glands and ovaries), as well as being responsible for the development of the male sex organs, testosterone has been linked to a range of behaviour and personality traits – typically, increased competition and aggression, with violent criminals often showing higher levels. More positively, testosterone increases energy, attention and confidence. In a series of experiments, Bernieri had participants (who'd had their testosterone levels measured) take part in a range of interactions, from introducing themselves to a camera to describing their personality to an interviewer. As expected, the subjects with higher testosterone displayed a more forward and confident manner, generally seeming more focused and expressive.[41] Could this shed light on the handshake studies from earlier? Putting cultural variations aside, maybe it's testosterone that partly accounts for the strength of our grip. With women having around levels five times lower than those of men, it would certainly help to explain why theirs is generally weaker. Equally, it would make sense of our hunch that people with bone-crushers tend to be alpha-male types.

Again, there's more than the whiff of reductionism here, but it does appear that, for better or worse, our testosterone levels partly define our first moments of interaction (and first impressions), particularly in high-stakes situations such as a job interview or first date. All of which seems grossly unfair to those of us who have low or even high levels. For starters, women are at an immediate disadvantage. But this is where the recent research of a team of social psychologists from Harvard and Columbia universities comes in, giving us one of the most prominent, and potentially life-changing, ideas to come out of the field: power posing. Seeing how alpha-male types tend to adopt expansive poses, whether a silverback beating his chest or company boss reclining in his chair, the basic theory is that our body language doesn't just reflect the way we feel but shapes it. Recruiting student volunteers, the team found that those who adopted these kinds of 'power poses' for a minute were more likely to gamble in a game than those who held more constricted poses. From saliva samples, it seemed that the reason was that those who held the power poses showed higher levels of testosterone as a result.[42] All of which leads to the heartening conclusion that we can indeed change our body chemistry – and thereby how we feel – simply by altering our body positions.

Unsurprisingly, the notion of power posing has attracted much media hype. One of the team, Amy Cuddy, went on to give the second-most watched TED talk and write a best-seller championing the idea. She recommends that just before an interview, important meeting or any moment in life when we need a confidence boost, we hold a power pose for two minutes. Many followers report big results, even saying that they get to feel what it's like to be US president. If nothing else, a bit of power posing might make your handshake stronger.

But, if all this sounds too good to be true, sadly, you might be right. More recently, power posing has fallen foul to what can be the scientists' worst nightmare: replication. Other researchers have failed to get the same results, finding that power posing had no real effect on hormonal levels.[43] So are Cuddy's fans making it up or, like the guy who poured OxyLuv into his coffee, simply experiencing the placebo effect? Perhaps. But, before we dismiss power posing altogether and go back to slouching in our chairs, volunteers who adopt the poses nevertheless report feeling more powerful. So while the results are inconclusive and the psychologists fight it out, it could still be worth a try.

Whatever the truth about power posing, for Cuddy all this relates back to her work on how we judge and influence people. Building on Asch's analysis, she suggests that one of the main dimensions we assess is how figuratively warm or cold someone is, associating warmth with a constellation of personality traits, including helpfulness, friendliness and, crucially, trustworthiness.[44] Intuitively, this makes sense, with the metaphor relating to our experience of physical warmth, perhaps even to our time in the womb, which we associate with a sense of security and attachment. Moreover, when it comes to our social judgements, it turns out that the experience of tactile warmth can activate feelings of interpersonal warmth. When a team of social psychologists investigated, they found that when volunteers were asked to judge a fictional character, their assessments seemed to vary according to whether they'd just been holding hot or iced coffee (they had been deliberately primed).[45] So while there's not a lot we can do about our body temperature or the weather outside, it could be a good idea to put that cold beer aside before shaking hands with someone for the first time.

———

Our greetings often engage another sense that we tend to forget when it comes to studying human interaction: smell. Whereas vision is our dominant sense and touch the first to develop, smell is actually the oldest. Otherwise known as the olfactory system (and including taste), it's our means of detecting the chemical composition of the world around us, something that even our earliest single-celled ancestors could do. In humans, the olfactory system most obviously consists of the nose, which is full of special neurons with odour receptors. While we use our smell to help gauge what's tasty or best avoided, it can also be a powerful trigger of memories and emotions. One of my friends came up with a word for it: 'memorific'. The likely reason is that the olfactory cortex sits close to the hippocampus, which is involved in processing and packaging our memories. It's why even neuroscientists recommend changing your perfume before going on holiday.

But it turns out that our olfactory system may also be crucial for communication – especially when it comes to our greetings. We see it most clearly in other species. Much to our embarrassment and disgust, we've all watched as a dog or cat goes straight for the backside of another, greeting them with a determined sniff. We might pull them apart, but this habit serves a crucial function. It's not so much that they enjoy the smell, but that underneath the anus there are two glands or 'anal sacs' that produce a key component of their communication: pheromones. These are scented hormones, usually carried in sweat and urine, that convey a whole range of information or chemical messages, cutting out the need for small talk. Like many mammals, our dogs and cats can tell each other's sex and availability, the state of their immune system, what they've been eating, even where they've been and what kind of mood they're in. They're also used to mark territory and each other, with every animal having a unique odour or biomarker. We all know about dogs marking their patch, but when bushbabies meet, they even spray each other – what's known as 'contact urine'. When cats rub noses or bump heads, they are doing something similar: passing on pheromones from glands around their mouths and foreheads (it's the other half of the answer to my mum's question about why Alice greets her with a kiss each morning). For such a solitary species, these chemical signals are an important means of managing relations, establishing who's who and forging bonds. It's why cat experts tell us that the best way to greet them is to hold out a finger – so that that they can first give us a good sniff.

Of course, we humans do not go in for such overt olfactory investigation, finding it all a bit uncivilised. Like most primates, over the course of our evolution we've gradually lost our sense of smell at the expense of other senses. Moreover, we tend to mask our natural odours with clothes and perfumes, duping each other with our Lynx Africa and Chanel No. 5. Yet, underneath all this, we also each have a unique scent and continue to secrete pheromones from various glands. Moreover, scientists have found that we too can communicate with our noses. For example, when volunteers were asked to sniff patches that had been worn under peoples' armpits while they'd been watching different films, they were able to identify (at least better than chance) whether they'd been watching a comedy or a thriller.[46] And it seems that our pheromones can also pass on emotions, with a similar experiment showing how participants' expressions tended to match the emotion they were smelling.[47] Not only can we smell fear, we feel it too. It's got scientists wondering how many other emotions we transmit through our odour – happiness, anger, attraction, love?

All of which suggests that our greetings, which are often the only time we make physical contact with each other, may serve an extra function. It would certainly help to explain the likes of nose-rubbing and cheek-sniffing, but maybe kissing and hugging too. A team from the Weizmann Institute of Science in Israel even wondered if the same could be said for our most common greeting (the handshake), noting how frequently we sniff our hands. To find out, they first established whether the gesture transmits the relevant chemicals, shaking hands with volunteers while wearing rubber gloves and testing them for various molecules. With the results coming back positive, they invited participants to their lab and secretly filmed them greeting the experimenter, who either shook their hand or didn't. As predicted, they found that the participants who received handshakes sniffed their right hands more afterwards (though there was some variation across genders). All of which led the researchers to conclude that an important function of our handshakes, albeit subliminally, is indeed to pass on our chemical signals.[48]

Perhaps the bigger point here is how much of what goes on in our interactions we're unaware of. So when we resort to that stock get-out phrase 'the chemistry just wasn't right' to explain why we don't want a second date, maybe there is some truth to it. However disconcerting, as Frith explained, we're not conscious of most of what's going on in our brains: 'It's where the gut feeling comes from.'

———

It's clear, then, that the way we approach and experience our greetings is shaped by our physiology, however much our rituals vary from person to person and culture to culture. At some level, as humans, we all share the same basic biological make-up. Yet it's also clear that all this differs with age and gender. And, across cultures, we see differences in how children and adults, and men and women, behave socially, which is evident in our greetings. So is this also a matter of differences in our brains and bodies? Of course, all of this touches on contentious issues.

But, where greetings are concerned, there are some factors that we can at least shed light on, if not settle.

As much as you might be able to coax a three-year-old into a handshake and hello, they'll never ask 'How are you?' or 'What've you been up to recently?' In September 2016, the Canadian prime minister Justin Trudeau knelt opposite Prince George to say hello. First, he tried a 'low five', then a high five and finally a handshake before giving up and doing his best not to look hurt. To some extent, all this came down to the fact that, like any three-year-old, George hadn't yet learned or got used to the conventions of normal social life. But it also reflects a more fundamental difference: up until the age of four, we're basically self-centred, unable to think about what others are feeling. As psychologists have shown, using the Sally–Anne test, we lack a theory of mind. It's only with changes in our mPFC that we start to appreciate that other people have thoughts and feelings too.[49]

Perhaps the biggest and most disruptive changes, though, come with our passage to adulthood. Again, the way we treat this and cope varies between cultures, but as a rule adolescence brings changing behaviour patterns, including increased sociability, a propensity to do risky or stupid things and a growing self-awareness.[50] Some of this comes down to physical changes brought on by puberty. But our brains are changing too. In particular, the prefrontal cortex (especially the mPFC) undergoes a substantial structural and functional development, with a sudden growth and reconnection of neurons, giving us our heightened consciousness and theory of mind. Hence, we become more self-aware and attuned to what other people are thinking, especially about ourselves. We blush more and feel awkward, worried that we're making an idiot of ourselves, especially around members of the opposite sex.

However painful all this is, these changes make a lot of sense. As we reach sexual maturity and leave the protective clutch of our parents, it's important that we're able to build our own social networks. So it makes sense that we gain the extra brainpower needed to consider what other people are thinking and feeling. We become conscious of the need to fit in, to adopt the styles, customs and manners (the etiquette) of our group. And, with our peer relations being so important, we're compelled to use the various bonding rituals that were once reserved for our mothers. As awkward as they are, they start to have real social value.

Thankfully, by our early twenties, our brains and bodies begin to settle down. Yet however much life wears at the edges of our self-consciousness, as we know we're never entirely freed from awkwardness and misunderstanding. Undoubtedly, a particular danger area is greetings between members of the opposite sex. Although highly ritualised, our customs often cross physical boundaries that are mostly reserved for families and partners. It's one of the reasons that hugging was banned in some schools in the United States: the teenage boys were treating them as a chance to feel the girls' breasts. And it's why the 'A-frame' hug is the norm among friends. But even in France, where cheek-kissing is standard, some admit that it's an easy way of flirting – a mini courtship. It has even led commentators (especially in the UK) to suggest that social kissing is a means of containing our repressed sexual desires, which can only add to our concerns about being misunderstood.

Compounding all this, greetings differ between men and women more generally. So, across cultures, men are more likely to use some kind of 'mock attack', whether a spear-throwing gesture or a fist bump. And even when we use the same rituals, we perform them differently, with men's handshakes tending to be firmer and their hugs involving extra slaps to the back. Of course, a significant part of this has to do with culture, being connected to much bigger social and political forces. You only have to look at our recent history to see how much things have changed: with the march to equality, so our greetings have converged. Yet differences remain and, arguably, always will, whatever the cultural changes.

It's a tricky issue, but as we've looked at our greetings we've touched on factors that point to some innate differences. Beyond the obvious bodily distinctions, we differ in our chemical make-up, with men producing more testosterone and women producing more oxytocin. As we know, these hormones have been linked to different kinds of behaviour: crudely, testosterone is associated with aggression and competition, and oxytocin (the 'hugging hormone') with bonding and trust. Moreover, some scientists have suggested that this has resulted in our brains developing slightly differently, with higher levels of testosterone leading to faster growth of the right hemisphere. And, from differences in our brains flow differences in our behaviour. So, from birth, girls are more attuned to human faces and sensitive to body language, developing a theory of mind earlier, while boys are more attracted to inanimate objects and lining things up. Baron-Cohen has put it starkly, suggesting that the female brain is predominantly hard-wired for empathy, while the male brain is predominantly hard-wired for understanding and building systems.[51]

For evolutionary psychologists, all of this makes sense, given the basic division of labour between men and women over the past 200,000 years or more. For others, it's the worst kind of bio-logical determinism, holding up social progress and justifying the continued marginalisation of women in public life. Whatever our biological differences, the main point is that we've learned to override them. Yet it's clear that between the extremes – and as our greetings suggest – our physiological make-up does come into the mix. In short, we might not be bound by our biology, but we are at least primed by it.

So what of my hug with Amma? As yet, I haven't noticed any radical changes in my life – my back is still playing up – and remain unconvinced that she has any special powers. As a friend said when I mentioned that I'd been to see her: 'Well, you'd expect to cure *someone* if you've hugged that many people.' In the end, her healing powers might well come down to the law of averages: eventually, there's going to be someone who feels better after seeing her. I can't say I felt any great spiritual awakening either – more a sense of relief that I hadn't been caught pushing in and could finally head home. That said, my hug was actually a nice experience, even comforting, which I guess is something, coming from an Englishman. Maybe it was Amma's unconditional love flowing strong, or maybe it was the extra shot of oxytocin and dopamine. As I drove home, though, trying to make sense of my increased feeling of well-being, I realised that it wasn't so much about the hug, but the experience around it. After days holed up writing this book, it was just good to have social contact again, to sit down with strangers and talk about what made them tick, to meet someone who had spent her life connecting people. My social brain was purring. And the next day, when I put my jumper on, I got that smell of incense and baby lotion.

8

Forget about Chimpanzees

On 16 January 2011, a small crowd gathered in a corner of Times Square in New York to witness the start of a unique event: an attempt to break the world record for the longest handshake. The field had been whittled down to four teams: two from the United States, one from New Zealand and one from Nepal. With the light fading, the competitors took their grips and began shaking. After fifteen minutes, one of the American teams realised they weren't up to it and dropped out, wishing the others luck. The rest kept going, aiming to break the previous record of twenty hours. But eight hours in, with temperatures dropping to minus eight, one of the remaining Americans collapsed. While paramedics attended the scene, the others kept going. They had form. One of the New Zealanders, Alastair Galpin, had held the original record and broken many others, including the longest distance for throwing a light bulb (28.7 metres) and the fastest time to peel a hard-boiled egg (5.09 seconds). He'd been practising for this event by shaking a bottle of sandwich spread – for 165 hours. For their part, as well as holding the current handshaking record, one half of the Nepalese team of brothers had given the most kisses in one minute (116). They shook on through the night and into the next day, breaking the twenty-hour mark. Thirty hours in, despite feeling cold, tired and nauseous, it was clear that neither team was going to give up, having exceeded the expectations of the event organisers, who now wanted to go home. With flights to catch, after thirty-three hours and three minutes, the teams decided to call it a draw, simultaneously letting go. Galpin rushed to the nearest toilet, while his partner hobbled to bed in silence.

There's no doubt: humans are odd. Two years after the New York event, a pair of actors from Turkey and Armenia again withstood freezing temperatures, this time in Georgia, shaking hands for a full forty-three hours, hoping to improve relations between their countries. While there's no official record, in 2016 two baseball players from the Coastal Carolina Chanticleers performed a handshake that had between thirty-eight and forty-five different moves. Records aside, as we saw in Chapter 4, humans have an endless array of weird and wacky greetings. So, having set out the evolutionary origins of our greetings and their physiological underpinnings, how come they vary so much? Looking at our closest relatives, we might know where kissing came from, but how is it that some of us became double- and triple-kissers, while others abandoned kissing altogether? How did we come to devise such elaborate routines and use our gestures to set world records and symbolise global peace? As we'll see, in many ways, this is the story of what makes us unique, even special, how we really did break or at least stretch our ties with the rest of the animal world.

To get some insight into how humans became so diverse, I arranged to meet Professor Robert Foley, co-founder of the Leverhulme Centre for Human Evolutionary Studies at the University of Cambridge. Having spent chunks of his career foraging about for human remains in East Africa, Foley is renowned for his theories on the evolution of cultural diversity. From the photos on the centre's website, showing him in various dusty places, dressed in khakis and a wide-brimmed hat,

I'd imagined a real-life Indiana Jones. But as he appeared in the usual academic get-up of jacket and suit trousers, he explained that he was still catching up on various admin, having come back from fieldwork in northern Kenya, where he'd been working on a new project on the origins of modern humans. Foley had originally set up the centre with Marta Mirazon Lahr, his wife and also an evolutionary anthropologist at Cambridge, as a place where different disciplines could come together to investigate the human story. Researchers now come from everywhere. 'It's like a mini UN,' he said. 'It's a nightmare knowing how to greet everyone. Two kisses with the French. The Italians start left...'

I began to explain how, having delved into the animal world, I was hoping to trace the evolution of our greetings through various species of human, which I still struggled to pronounce.

Foley looked doubtful. 'Forget about chimpanzees, *Australopithecus* and all that lot,' he said, pulling a pad from his desk.

'Really?' Suddenly, I doubted my thinking. But, if I could short-cut 6 million years of history, so much the better.

Foley explained that while we might have inherited a propensity to greet, when it comes to tracing the diversity of our rituals, we only need to go back 200,000 years to the first *Homo sapiens*. He opened his pad and drew a diagram of a tree with branches splitting off. Rather than different species, they represented populations of humans, along with their various customs. The point is that human behaviour has its own evolutionary tree, growing much faster than the rate at which different species emerge. So, starting from a small population somewhere in East Africa, human culture broke off along many paths. It's a paradox that distinguishes our species: low biological diversity (genetically, we're pretty much all the same) but high cultural diversity. To some extent, it comes down to our wiring (or lack of it). Unlike other species, we come into the world with little behaviour hard-wired. For the first years of our life, our neurons remain relatively unconnected, which means that, as well as being largely helpless, we are also uniquely malleable. Combined with the size of our brains, this means that we have an unrivalled ability to adapt and innovate. We don't have to wait for genetic mutations to shape our behaviour, but can pass it on from person to person, group to group. At its most basic level, this is how Foley and others who study human evolution define culture: as 'socially transmissible behaviours'.

To be sure, other animals show a capacity for culture. As Foley described, whales are a particularly interesting case, having sonar calls that vary from community to community, almost like accents. Unsurprisingly, our closest relatives demonstrate some of the most diverse behaviour. At the last count, primatologists compiled thirty-nine different behaviour patterns that vary from one chimpanzee community to another, including various kinds of tool-making and grooming.[1] More behaviours are being discovered all the time. In fact, Foley had just been sent a new paper by a colleague at Harvard detailing how high-arm grooming, which involves a pair of chimpanzees each clasping one hand above its head and using the other to comb through the other's fur, has emerged and spread through different communities as a result of social learning.[2] But as Foley puts it: 'The odd variation in chimpanzee handshake does not compete with more than 6,000 different ways to say hello.'[3] Moreover, a defining aspect of human culture is that it's cumulative. That is, one innovation builds on another, with each generation adding to the last, so that human invention is endless. So while chimpanzees have been stuck cracking nuts with rocks and grasping hands over the past few million years, we humans have gone on to make axes, send rockets into space and devise entire handshake routines.

As Foley concludes, 'culture produces cultures.' But the core of his work has been explaining why we get such variation. In many ways, his theory, which he first developed in the 1980s, is straightforward: the key factor is ecology – the environment around us. It goes back to the basic fact that, with our over-sized, gas-guzzling brains, we were constantly on the search for food. So, starting from a small pocket in East Africa, as the human population grew, it broke off in different directions, first spreading across other parts of the continent. Then, sometime around 65,000 years ago, an intrepid group made their way out of Africa, with populations making it to Europe around 50,000 years ago, and eventually colonising every continent (bar Antarctica) by 15,000 years ago. At a talk afterwards, someone asked why we were in such a hurry. Foley half joked that it may have also had something to do with younger males wanting to get away from the older men monopolising all the women.

Inevitably, as the different populations adapted to their sur-roundings, they developed distinct ways of living. So those who braved the icy climes in the far north of Asia and the Americas became seal hunters, wore fur and lived in snow houses, while those in the tropical rainforests foraged for fruit and could get away with fewer clothes. That much is obvious (and takes us back to the thinking of Franz Boas, whom we met in Chapter 5). Yet as Foley and other paleoanthropologists recognise, human culture is more whimsical than that. Take the Lower Omo Valley in Ethiopia, for instance, which is close to where the oldest remains of *Homo sapiens* were found. Here, within a small area, you can still encounter sixteen distinct ethnic groups, many of which have their own unique language, dress and lifestyle. The

Hamer, for example, are considered masters of body decoration, binding their hair with ochre and wearing intricate jewellery, while the Mursi, just across the river, are famous for their fierce stick-fighting and for fixing 10-inch clay plates in their lips. It's the same closer to home, with accents and habits varying from region to region, from one valley to the next.

The defining thing about much of human culture, then, is that it has no clear function. In a word, it's symbolic. At the most basic level, we're able to take something and make it stand for something else: gestures and words, for example. So a thumbs-up and 'hello' convey a meaning that has no direct association with the action. More profoundly, we get art and storytelling. Archaeologists are continually revising their view on when this capacity for symbolic culture evolved as they make new finds. Recently, shell beads and a small piece of red ochre carved with a diagonal pattern found in a cave in South Africa suggest that it had developed by 80,000 years ago, while human burials from around 100,000 years ago may represent a form of ritual, possibly indicating a belief in the afterlife. Yet, in all likelihood, it goes back much earlier, to a change in our brains when, for whatever reason, our consciousness upped a level: we became conscious of being conscious. We could now question things, like why we're here, and imagine a world beyond our own. And so, as we came up with answers, we can see the origins of religion and other beliefs. We were no longer bound by our physical world, but had become creatures of culture.

Bringing all this back to greetings, for Foley, a key dimension of the early human experience is that we lived in fission–fusion communities. So while we might have belonged to extended kin networks, perhaps having a residential group of twenty or so and being part of a bigger language community or tribe of around

500, as we searched for food we would have often broken off into smaller groups or bands, consisting of just a few people. Sometimes we wouldn't see the others for days, weeks or even months. So it was important to have a means of recognising each other, of distinguishing ourselves as one of the group – 'us' not 'them'. It's a big part of why we get all the different forms of body decoration, funky hairstyles, clothes that go beyond being merely functional, strange accents and other weird behaviour. As the historian Michael Cook puts it: 'humans have a quite remarkable capacity to tie themselves into knots by devising elaborate and ultimately arbitrary rules.'[4] In many ways, this is the point: what better way to demonstrate that we're one of the group than by conforming to customs that make no sense? And so it is that we get many of our oddest greetings. How else can we explain the ritual in Papua New Guinea of squeezing your nose shut with your left hand while pointing to your naval with the right, or the fact that France has a kissing map and the Freemasons are rumoured to have so many secret handshakes? Beyond our expressions of pleasure and testing our relationships, we're demonstrating that we're one of the gang.

As Foley explained, a key aspect of all this is what happens at the boundaries between groups or cultures. As we saw in Chapter 6, chimpanzee communities tend to attack each other while bonobos have sex. Falling somewhere in between, humans veer between competition and cooperation. On the one hand, early human groups would have competed for resources, but on the other they would have exchanged goods and, to keep the genes mixed up, mature females. So, as Foley speculated, it's likely that they had certain 'iconic greetings' that were recognised across communities, such as the wave or the handshake. These were the earliest symbols of multicultural living.

An exception, they say, proves the rule. As our meeting came to an end, Foley mentioned the Turkana people of northern Kenya who, as far as he knew, often didn't use any greetings at all. In his experience, they just came and sat next to each other without any extra complication.

———

For the vast majority of our history, this is how we lived: moving around in bands and tribes made up of extended kin, living hand to mouth, engaging in sporadic cycles of conflict and cooperation, constantly splitting off in different directions and forming new cultures. But then, around 12,000 years ago, coinciding with a warming up of the planet, something happened that would dramatically change the course of human history: we began to grow stuff and settle down. Starting somewhere in today's Syria or Iraq and eventually spreading or springing up across the world, the first agricultural revolution fundamentally changed the way we live. Producing our own food and living in one place meant that we could accumulate resources. We could feed more children and own stuff. Inevitably, our fixed communities grew, absorbing surrounding tribes. Hamlets became villages and tribes became chiefdoms. By the fourth millennium BC, the first towns and cities appeared in Mesopotamia. People were now living in densely populated areas and settlements of 50,000 or more. Across the world, communities became civilisations and cultures became mega cultures. Their interaction has been the major story of the past few thousand years, with successive kingdoms and empires coming and going, swallowing up different peoples and parts of the world. Day to day, humans found themselves living in ever bigger and more complex societies. We have had to adapt our behaviour to a new kind of ecology – one made by us.

So how did all this affect the way we interact? And how have our greeting rituals changed over time? While I originally trained as a historian, I realised that answering these questions would require a different approach to the one I was used to. Whereas my own research focused on the history of international relations, using official documents to piece together decision-making – what was once regarded as the standard practice of historians – this is the history of the everyday, as much about the ordinary person as world leaders.

A pioneer of this kind of history has been Sir Keith Thomas at the University of Oxford, widely regarded as one of the leading social historians of the twentieth century. Starting out in the 1950s, at a time when diplomatic and political history was still the vogue, Thomas turned his attention to more personal matters. Inspired by the approach of anthropology, he wanted to get inside the minds of ordinary people, to understand how they lived and what motivated them. It meant looking at things like people's attitudes towards religion, the changing role of the family, how people treated animals, their hopes and desires. While Thomas's own work focused on early modern England, his approach has inspired historians more generally to look at all aspects of behaviour, even things like daily habits and rituals. In his introductory remarks at a conference on the history of gesture, Thomas set out the scale of ambition that's possible when looking at a seemingly trivial aspect of life: 'To interpret and account for a gesture is to unlock the whole social and cultural system of which it is a part.'[5]

Now eighty-four, Thomas is currently writing a history of manners. When I got in touch, he said that he would be fascinated to hear about my own project, even inviting me for lunch at his college, All Souls – though he did mention that he wouldn't be able to share all of his references. I'd read that Thomas had recently taken issue with popular historians, describing a kind of 'parasitic' relationship in which they steal all the research and ideas of proper academics – which is exactly what I was hoping to do.

Having originally won a scholarship to Balliol College, Thomas has a reputation for being the 'most brilliant mind in Oxford', which, by extension, would make him just about the cleverest person in the country. So as he appeared from one of the Gothic arches, I was relieved to find that he was immediately warm and friendly, asking about my journey and complaining about the rain.

Thomas took us through to the buttery, where lunch was being served and he was greeted with a 'Good afternoon, Sir Keith' by the various staff. 'For better or worse, I've been here since 1952,' he said. He comes from a small village in Wales, though he has lost all trace of an accent. When he first arrived in Oxford it was a rule that students didn't shake hands with the fellows. Like the high tables, it set them apart.

We were joined by a couple of other fellows, whom I assumed were also geniuses of some sort. 'Please meet my guest, Andy Scott, who is writing a book on greetings,' Thomas said. I felt a bit embarrassed. It didn't exactly sound heavyweight.

'How do you do,' said the woman, holding out her hand. I wasn't sure if she was being ironic, but guessed that certain phrases survived in Oxford. All in all, I felt like I was back under the gaze of William Hanson, as I tried to work out which bit of cutlery to use.

After lunch, we retired for tea in one of the college's common rooms, complete with leather armchairs and lined with old photo portraits. 'It is rather like a mausoleum,' Thomas smiled, before going through some of the faces, pointing out famous writers, philosophers and politicians. Eventually, I stopped asking who they were and nodded along.

Thomas sat back in his chair, stirring his tea. 'Well, fire away.'

I was hoping to get through around 12,000 years of history, so it was difficult to know where to start. I ended up quoting back to him his comment about gestures unlocking entire social systems, saying that I hoped to do the same with greetings.

'Ah, yes – well, easier said than done, of course,' he smiled.

It led me to ask how we can hope to really know how the average person lived and behaved hundreds of years ago. As one of Thomas's own role models, the anthropologist Edward Evans-Pritchard, had put it: 'How can an Oxford don work himself into the mind of a serf of Louis the Pious?'[6]

'Basically, you need to read everything,' Thomas said. Diaries, letters, poems, plays, paintings – they all offered clues to the texture of daily life. Famously, Thomas has a particular method for gathering evidence. Whatever the source, he takes notes on a single side of paper and then cuts them up, quote by quote, filing them in envelopes according to different themes, such as witchcraft, family life or pets. He has hundreds stashed away. Every now and then, he'll pick one out, empty its contents and hope that some order emerges. It has its problems, though, as he's forever finding stray slips behind a cupboard or in the garden, sometimes fluttering away – an idea lost forever. But, like an anthropologist, Thomas doesn't start writing until he feels that he can inhabit the world of his subjects.

It still seems to me, though, that the historian's perspective inevitably falls on a particular section of society: those who are literate and bother to write things down. Part of it goes back to Norbert Elias, whom we met in Chapter 3, and his theory of manners – that they grew out of the rise of the nation state and its need to control violence. From Elias, we get a sense that our manners and rituals came from the top, trickling down from the royal courts, civilising the masses. While there's obviously truth to this, it's clear that it's not the whole story. For instance, it's thought that cheek-kissing in France – and all the problems that have gone with it – originated as a peasant custom, which was gradually adopted by the urban elite through migration.[7] Moreover, taking the longer view – by a few thousand years or so – my own theory was that our manners must have gone through a transformation with the first agricultural revolution, as we came to live in bigger and bigger communities, often along-side strangers. Surely it was in those first cities in Mesopotamia that we became accustomed to what Goffman called 'civil inat-tention'. To cope with living among so many unfamiliar faces and manage our day-to-day encounters, we would have relied on common and recognisable behaviours and, increasingly, all those fleeting gestures we use to acknowledge each other's presence, such as a half-smile or a flicker of eye contact. With so many people around, we'd have sometimes resorted to speedier greetings, so a quick smile, nod, 'Hi' or rhetorical 'How are you?' would have been enough.

'It's a reasonable guess,' Thomas said. But he went on to explain that, from his own work on early modern England, it was clear that visitors to London were struck by the way that people didn't acknowledge each other in the street, not even with a glance. 'It really wasn't very civil at all.' Even in the eighteenth century, the English were getting a reputation.

For Thomas, the big driver of manners and civility was our increasing interdependence, particularly through commerce.

As we traded with each other, we relied on certain behaviours to smooth our transactions. Despite having one of the sharpest minds, Thomas has never shied away from common sense. From my own studies, it's a major theme of the last few thousand years: with advances in transport and as various empires expanded, the world became ever more interconnected, until it became one giant web.[8] It stands to reason that we developed – sometimes imposed – certain common behaviours. It also helps to explain why the world of diplomacy can seem so staged and stuffy. Going back to when the first nation states emerged and interacted in Renaissance Europe, it was important to have a set of rituals and standards – what's known as protocol – to smooth international relations and insulate them from changes in government, as well as the whims and gaffes of individual leaders. But again, thinking of my discussion with Foley and those first tribes on the savannah plains, all this goes back much further.

When I explained that I'd been delving much deeper into the past, searching for the evolutionary origins of our greetings, Thomas was interested, but sceptical about stretching our links with the animal world too far.

'It's a rather banal thing to say, but I see the last two thousand years or so as the age of culture – when culture consciously over-rode nature,' he said. 'Of course, we have these primal instincts, which we can't abolish,' he conceded, 'but we can regulate them.'

I double-checked when the age of culture began. 'Well, I suppose you could go back further. You'd need to include the Greeks,' he said, trailing off.

All this took me back to a core tension that had been running through my research: the extent to which human behaviour is a matter of nature or culture, innate or learned. Although the debate is less polarised and angry than in the days when Thomas was starting out, academics are still divided, often talking very different languages. While I didn't challenge Thomas at the time, the more I looked into our greetings, the more I realised that it was something of an artificial distinction, with no clear divisions or turning points. It partly depends on how you define it, but from those patterned bits of ochre and shell beads found in South Africa, it's clear that the age of culture is much older. The thing that's characterised the human story is our accumulation of culture, generation on generation – something that's been enabled by our unique brains (our nature).

But if we are to talk about transitions, we surely need to go back to the first agricultural revolution, when we started to live in ever-expanding communities – including far more people than the 150 that Robin Dunbar has suggested our brains were adapted for (see Chapter 6). For Dunbar, along with the likes of the Israeli historian Yuval Noah Harari, the key to understanding how we coped in these growing communities, living and cooperating with unrelated strangers, is culture. Specifically, it's been our tendency to come up with stories about ourselves, or what academics call 'imagined realities', that has bound us together in bigger and bigger groups. Most powerful has been religion, but there's also been our political ideologies and national identities – and, of course, football. These are causes that we have fought and died for. Yet, arguably, they are all made up.

Where Thomas and the social historians come back in is by showing how our everyday behaviours have reflected these bigger cultural constructions and changed over time. So when it comes to greetings, we can see how our various rituals have in effect been co-opted by our cultures, signifying our broader attachments and beliefs. Whatever the evolutionary origins of the handshake, this helps to explain why some of the earliest depictions in Greek and Roman art are interpreted as symbols connecting people to the afterlife. It's the same with the kiss. Rather than simply expressing our relationships, it became embroiled with religion, symbolising the transfer of the spirit. In his Epistle to the Romans, St Paul instructed followers to 'salute one another with a holy kiss'. And so the 'holy kiss' became a common greeting among early Christians and a central part of Catholic ceremony.[9] And, whatever the evolutionary origins of the *hongi* – whether or not it started as a method of olfactory investigation – Maoris describe how it too signifies an exchange of spirits, recalling the legend that the first woman was created in clay, only coming to life when the god Tāne breathed into her nostrils.

In the extreme, there's the most notorious greeting of all: the Nazi salute. Known in German as the *Hitlergruß*, literally meaning 'Hitler greeting', it was borrowed from the Italian Fascist Party which, in turn, believed it was an old Roman ritual. With no hint of irony, Hitler himself recalled how it must have come from an ancient sign of peace.[10] Of course, far from being a peaceful gesture, it was instituted as a sign of people's commitment and loyalty to the Nazi regime, eventually being made compulsory in 1933. Everyone was required to do it, from shopkeepers – who greeted their customers with '*Heil* Hitler, how may I help you?' – to children at school. There were even signs on telephone poles and street lamps reminding people of their duty. As their huge rallies demonstrated, the Nazis were determined to create the appearance of mass support and enthusiasm. While the Hitler salute was made illegal in Germany after the Second World War, the communist regime in East Germany demanded new gestures of obedience and conformity. And among socialist leaders, the kiss and embrace became the fraternal kiss and embrace, demonstrating the special connection between socialist states. Famously, there is the image on the Berlin Wall of Soviet leader Leonid Brezhnev planting one directly on the lips of his East German counterpart Erich Honecker when they met in October 1979. During Communist rule in Czechoslovakia, the proper greeting was Čest práci ('there's honour in work'). The point is that our greetings have not just been personal gestures, but public expressions of our commitment to a higher being or cause, blurring our evolutionary and cultural history.

As Thomas explained, though, greetings are essentially an 'expression of our social relations'. The rituals indicate our relative status. As we've seen, it's true across the animal world, with different species using various gestures of submission and dominance. But humans are different, having created ever more complex hierarchies. Again, it goes back to the first agricultural revolution, when our communities grew and we started to accumulate resources. Whereas there had always been big men or alpha males, who would have come and gone according to their physical and mental qualities, they now had more formal responsibilities, such as protecting and distributing the excess food supply, and would have recruited supporters. And so we see the origins of the state, with a ruling class passing its privileges along family lines. As communities grew, the big men became chiefs, kings and emperors. They wore elaborate headgear, sat on thrones and demanded ritual displays of deference. Among the tribes of Hawaii, for example, when commoners encountered a chief, they were required to prostrate themselves. It doesn't take much to draw a direct line to the royal courts of Europe or Asia, with all their stylised bowing, scraping and kowtowing. In the sixteenth century, for example, Emperor Akbar of India was credited with inventing the *kornish*, which involved a courtier bending at the waist, touching his forehead with his right palm and then sweeping his hand downwards in an arc until he almost touched his feet, and then repeating it twice.[11]

Moreover, as rulers tried to control vast territories and diverse populations, they deepened hierarchies, creating extra tiers in society, which in turn demanded displays of deference from those below. As Thomas described, in early modern England these codes of courtesy were universal, with everyone knowing their place. So in the medieval period, if a knight rode past, a serf would keel over, while in Tudor times people were forever doffing and donning their hats as a sign that they knew their rank. Failure to comply could land people at the whipping post or in the stocks.[12] Similarly, as part of the caste system in India, codes were developed to regulate interaction. In Amma's home region of Kerala, for example, those in the bottom caste ('the untouchables') such as her were required to keep sixty-four paces away from members of the highest caste, the Brahmins. Even in Sweden, long renowned for being a bastion of equality, during the eighteenth century people from lower orders were restricted from approaching their social betters.[13]

Conversely, as with our gestures as a whole, our greetings have been used to challenge authority, becoming little acts of protest, what some have called 'weapons of the weak'.[14] And so, for example, an oatmeal-maker brought before a royal court in early modern England refused to take his hat off to the judges, as did a Scottish soldier who was being tried at a royalist council in October 1645.[15] Beyond non-compliance, we've developed gestures of rebellion. Some have been more private, captured here by an Ethiopian proverb: 'When the great lord passes, the wise peasant bows deeply and silently farts.'[16] Or they've been more explicit, such as in the scene in *Braveheart* in which, standing on the battlefield before the English army, Wallace's men turn around, bend down and pull up their kilts, revealing all. There may be some artistic licence here, but mooning – the invitation to 'kiss my arse' – has been a tactic employed in conflicts throughout history, triggering feelings of embarrassment and anxiety among opponents. During the 1984 uprising against oil exploration in Nigeria, for example, Ogharef women pulled up their dresses, causing the security forces to run away, believing that they were being cursed.[17]

As Thomas described, though, perhaps the most important and radical group in all this are the Quakers. A close-knit Christian sect, mostly from England and Holland, they came to America as missionaries in the seventeenth century. Renowned for being devout and egalitarian, they were also famously rude. They refused to call people 'Your Grace', bow or take off their hats, maintaining that such acts represented the worship of man. Instead, and often attracting abuse, they developed their own, simpler gestural code, adopting the handshake as their preferred greeting. Significantly for us, it's suggested that it was the Quakers who popularised the handshake as a greeting: up until then it had been notably absent from etiquette guides.[18]

Being a symbol of equality, it's no accident that the spread of the handshake coincided with the rise of democratic ideals, which had sparked revolutions in North America and Europe. A prominent trendsetter was Thomas Jefferson, who, on becoming US president in 1801, did away with the bow as a greeting, opting for the handshake instead. Notably, it was Jefferson who penned the Declaration of Independence, with its famous statement that 'all men are created equal'. In reality, of course, despite the heady rhetoric and political upheavals of the time, Jefferson's phrase fell far short of the reality. Aside from the continuation of slavery, women were left out of the equation. It was only with the second wave of feminism in the 1960s and 1970s that social behaviour really changed. As women were recognised as equals in public life by law, the rules around 'ladies first' were increasingly seen as old-fashioned, even chauvinistic, while hand-kissing was replaced by the more equal cheek-kissing and handshake. Overall, then, looking across our history at the rise and fall of hierarchies, it's clear that these little rituals have been a barometer of big political and social changes.

———

Keen to bring the story up to date, I wanted to get Thomas's thoughts on what some commentators have characterised as the 'greetings inflation' of recent years, whereby handshakes have become hugs, hugs have become kisses, and so on. Even our written communications reflect the trend, with the exclamation mark in heavy use these days, so 'Hi, how's it going?' has become 'Hi! How's it going?!' or 'Hi!!! How's it going??!!!' Looking back, what makes it even more notable is that it represents something of a U-turn. As we know, for centuries etiquette guides stressed the importance of bodily discipline and restraint.

Having lived through most of the last century, Thomas was in no doubt that we've seen a dramatic change in recent decades. He described how not long ago he'd met the head of a university press in New York who'd kissed him on the lips. 'I was amazed,' he said, eyes lighting up. And he now found himself shaking hands or hugging where he wouldn't have before. But for Thomas, all this points to a bigger trend: what he characterises as 'the growing cult of informality and friendliness'. People no longer show inhibition about eating and swearing in the street or exposing their bodies and expressing their emotions in a physical form. 'Now when I walk into my bank, I am greeted with "Oh, hello, Keith". It's amazing how things have changed,' he said.

I took the chance to ask what the etiquette was with someone with his title. 'Please, call me Keith,' he said, making the point. 'It just adds another complication to your life that you don't want.'

Explaining all this is another matter, though. We could point to the influence of the media, with today's royal courts being found in Hollywood or Bollywood. As a rule, our actors and presenters are more expressive types. Watch any chat show and the guests are often hugging and kissing with abandon. Yet while they might lead the way, you only have to look at a similar show from a few decades ago to see that it's been a trend *within* the media too.

No doubt a big part of the answer goes back to the cultural revolutions of the 1960s and 1970s, with the effects still being played out today. In many ways, they marked a conscious break from the strict codes and certainties of the Victorian past. Rather than demonstrating how much we'd controlled our natural impulses, we now sought to emphasise that we were in touch with them. The liberation of sexual relations brought changes in other behaviour, such as dancing styles, while the relaxation in attitudes towards homosexuality saw people gradually shed some of their hang-ups. As Thomas has described, in the eighteenth century men seen kissing were likely to be accused of sodomy, with homosexual acts punishable by death from 1533 to 1861.[19] It's perhaps no wonder that, following Erasmus's earlier visit, it started to die out as a male–male greeting. Now, following the decriminalisation of homosexuality in Britain in 1967 and changes in public attitudes, hugging and kissing is coming back, even among heterosexual men.* Perhaps the main consequence, though, has been the changing role of women (and men) in society. Of course, women have had to do a lot of the adapting, as they've entered male-dominated environments, but the effect has worked both ways: our workplaces (and social places) have gradually become more feminised – less hierarchical and more empathetic (going by Simon Baron-Cohen's logic, discussed in Chapter 7). Moreover, as Western economies have moved from manufacturing to service industries, more interpersonal skills are required and some of the old ideals of manhood no longer have the same value or appeal.

For Thomas, a big factor in all this has been the decline in deference. As he explained, we still have our social hierarchies, but we live at a time when there is a 'general merito-cratic assumption'. It goes back to the Renaissance and the Enlightenment: the idea that it was virtue that mattered, not birth. Thomas suggested that, more recently, the Second World War marked something of a dividing line, throwing us together and breaking down social barriers. As he observed, one of the markers has been the decline in hat-wearing. Look at photos of football crowds from the 1930s and you'll see how everyone is wearing flat caps. From Tudor times, hats were used to greet other people, as a sign of respect. As social hierarchies and deference have waned since the war, so too has hat-wearing. 'I'll give you that one,' Thomas smiled.

More generally, the erosion of formal hierarchies and author-ity has meant that people have become more spontaneous and less buttoned-up. But, as Thomas described, there have been

---

* This line of reasoning becomes less convincing when you consider countries like Saudi Arabia and Iran, where men are more tactile with each other but homosexuality remains illegal.

problems with all this, especially for those who have lived through the transition.

'Every generation of history students tends to see their own era as transformative,' he said, 'but the speed of social change really is greater than ever.'

For the likes of Thomas, who grew up in more strait-laced times, our era of informality and friendliness has brought confusion, such as over knowing what kind of greeting to go for. Moreover, while this cultural revolution is still being played out, the pace of change has been uneven, with different generations and areas accustomed to different social standards and expectations. It's a form of culture clash, adding extra confusion to our interactions.

'It's really quite awkward,' Thomas said.

'I know,' I said.

———

For all that our greetings have reflected profound social changes, they have also been subject to more mundane factors. After all, most of us spend most of the time getting on with our daily lives, not trying to change the course of history. As Thomas recognised, one such dimension has been health and hygiene. For example, he wondered how far modern dentistry and breath-fresheners have been a precondition for the turn towards social kissing in modern times. Back when the Romans took it up, they perfumed their breath with myrrh. There have also been concerns that our physical greetings play a role in transmitting germs. In the first century AD, the Roman emperor Tiberius issued a decree banning lip-kissing, believing that it was responsible for spreading a fungoid disease that had disfigured the faces and bodies of nobles.[20] And it's been suggested that the decline in kissing in England was a consequence of the Great Plague of the mid-1660s.[21] More recently, a whole range of authorities recommended a ban on kissing in the wake of the 2009 H1N1 (swine flu) pandemic. Even in France, employees were asked to respect a one-yard person-to-person buffer and use a wave instead. All of which makes air-kissing seem less like a pointless affectation.

Yet it might be that the biggest villain is the handshake. Self-proclaimed germophobe Donald Trump famously described it as 'one of the curses of American society'. It turns out that he might be right, with scientists revealing that viruses spread more easily via the hands than via the lips. The authors of a recent study published in the *American Journal of Infection Control* recommended that people use the fist bump instead, which transmits one-twentieth of the bacteria of a handshake.[22] It seems that the message is getting through, with two in five Americans admitting that they have hesitated about shaking hands, while 15 per cent say they now use the fist bump for health reasons. So perhaps the Obamas weren't just trying to look cool (or being ironic?) when they famously fist-bumped, but setting a presidential example.

Finally, in all this there's a factor that's by definition hard to pin down, but undoubtedly important: fashion. Of course, fashion doesn't stand alone, but is shaped by social, economic and technological trends. Yet although fashion has its innovators, it also depends on that most human thing: conformity. So while the first to adopt a particular style may be making a social or political statement, the rest of us are often just fitting in, trying to be one of the gang or following the latest trend.

It's similar with our greetings. Consider the diagonal clasp or fist bump, for example, which have become increasingly popular. While their exact origins are unknown, they echo the rituals that were used by West Africans brought over to the Americas as slaves and then by black soldiers during the Vietnam War. What became known as 'the dap' (standing for 'dignity and pride') was a four-part handshake that signified black solidarity. But at a time when the Black Power movement and racial unrest were growing in America, it was seen as a form of insurrection and banned, with many soldiers court-martialled and jailed for using it. Like different jeans or hairstyles, variations of the dap, such as the diagonal clasp, became symbols of youth rebellion and anti-establishment attitudes, consciously adopted as alternatives to the handshake. Then they became cool and are now popular among white middle-class college students – even health-conscious scientists. Thomas was less sure, though. 'Really?' he said. 'People do that here?'

9

Back to the Future
or a Brave New World?

Knowing I was writing a book on greetings, a friend sent
me a joke. Two mind-readers meet backstage. One says
to the other, 'You're fine, how am I?' It might not be
side-splitting, but I found myself taking it seriously, as though it
were another example of cultural variation. I'd just been reading
about the very real potential for telepathy, so it didn't seem so far-
fetched. The only question was whether we'd keep up the phatic
communication (empty chat), which we discussed in Chapter 2.

Having examined the past and present, then, I was hoping
to take a look into the future, to explore how we'll be greeting
in the next decade, century, millennium even. As we've seen,
a problem with traditional etiquette guides is that they tend to
be of a particular time or, more to the point, stuck in the past,
whereas human behaviour is forever changing. I wondered
whether we can know what any of this will be like in tomorrow's world.

Watch anyone for long enough and an obvious point to con-
tend with is how technology is changing the way we interact.
So far, we've mainly looked at face-to-face greetings, but our
communications are more and more mediated through various
devices. First, in around 500 BC, came the letter, which was
followed by the telegram and telephone late in the nineteenth
century. And then we got email in the 1990s. Now we have
videoconferencing, Skype, FaceTime, WhatsApp, Snapchat and

whatever else, all of which we can access through a single device that fits in our pockets.

When it comes to greetings, in many ways nothing has changed – we've just adapted them to our various technologies. Generally, we use some version of hello, start off with a well-being enquiry and add a bit of small talk, maybe checking if the other person can hear or see us. We still vary our rituals according to the nature of our relationships – or at least that's what the etiquette advisors tell us to do. We still expect people to reciprocate and generally to try to match our gestures, whether through our tone, form of address or number of exclamation marks. We still look for extra clues to guide our impressions and work out what's on people's minds. And we can still get in a muddle over what gesture to use, agonising over whether it should be 'Dear', 'Hi', 'Hello' or 'Hey' – and so on. Given the speed and ease of our communications, 'Dear' can seem overly formal, but 'Hi' can feel a bit chummy, so sometimes we drop them altogether. But just as in our face-to-face interactions, our confusion also reflects the changing times – the trend towards informality and friendliness. You only have to look at the rise of the exclamation mark, emoticon and emoji or the corresponding decline in the use of capital letters to see how our use of technology mirrors deeper patterns.

Perhaps the real change in recent years is the *extent* to which our interactions are mediated through our devices. Just two decades ago, only a flash few owned a mobile phone. Today, it's around 60 per cent of us globally – more than the number of people who have access to a working toilet. By 2020, it's estimated that 70 per cent of us will have a smartphone, using it for most of our communications. Moreover, the internet is transforming the way we live. With the birth of social media, we don't just communicate online – we exist online. It's been the biggest migration in history, with nearly 2 billion of us now signed up to Facebook.

For many, this is the way forward. In a world where so many of us live far away from our friends and family, the internet has flattened physical barriers, allowing us to maintain our relationships and groups – wherever we are, the people we're closest to are only ever a click or a swipe away. Moreover, we can connect with people from across cultures, sharing our thoughts and ideas, breaking down ignorance and prejudice. In cyberspace, the world really can seem like a global village. Moreover, social media encourages us to be more open and share our thoughts: Facebook's 'What's on your mind?' is the perennial 'How are you?' And we don't need to worry about garbling our words, going red or any other awkwardness that comes with face-to-face encounters. The future is clear. As BlackBerry's slogan once put it: 'Always on, always connected'.

For others, all this is a bit alarming. Our gadgets aren't so much enhancing our interactions, but undermining them. One of the problems is that, as technology makes communication easier, we're drowning in it. Inevitably, it gets cheapened and some of our electronic communications have become more transactional. We've started to drop the little interactions rituals – to skip 'Hi' and 'How are you? and small talk. All of which can seem cold and jarring. As Randall Collins notes, these are the bits that build a sense of solidarity.[1] They may just be words on a screen, but they still trigger a response in our brains. In the end, the overriding point is that our mediated interactions are inferior. To varying degrees, we lose our physical presence and all the subtle cues and clues that are so important to our communication: eye contact, facial expressions, tone of voice, and so on.

As Susan Greenfield – the British neuroscientist and renowned worrier about the impact of technology on our brains – points out, these are the bits that can act as a brake, alerting us to when we're boring or offending someone.[2] It's why cyberbullying can be so extreme and Twitter exchanges can quickly get medieval. With anonymity, the barbarian comes out. Face to face, who would say what they've written on Donald Trump's Twitter feed? Online, we lose our civility, not even bothering with greetings (our acknowledgement of each other as social entities). Moreover, Greenfield warns that as we increasingly communicate through technology, some of our social skills will waste away: we'll get worse at reading people's body language and, saddest of all, lose our powers of empathy. As is often the case when it comes to the brain, we'll 'use it or lose it'. More generally, there's a feeling that our gadgets get in the way. We're forever checking our phones, looking at what our other friends are doing rather than engaging with the person in front of us. We're becoming 'pancake people', our relationships spread wide and thin, with no real substance. It has to be a bit sad when the Debrett's guide to 'netiquette' reminds us to 'indulge in face-to-face contact'.

It's hard to know where technology is taking us, whether it's to a brighter or a darker future. I can't remember why, but one of my bosses once used the analogy of boiling a frog. Apparently, as the temperature rises, frogs don't sense the danger, swimming around until it's too late. Could it be that we're doing something similar to ourselves? As someone who's still on his second Nokia (it's got to the point where it's now considered retro), I wanted to find out where things are heading, if there's a direction to all this and how it will affect our interactions. As we get on with our daily lives, most of us don't give it much thought, but there are some people who've made it their job to peer into the future – and I don't mean astrologers or people with crystal balls. Known as 'futurologists' or 'futurists', they keep pace with the latest technological trends, predicting where we'll end up, keeping us abreast of things like driverless cars, space travel and talking robots.

Ian Pearson, founder of the consultancy Futurizon, is one of the leading futurists in the UK. Originally trained as a physicist, he began his career working on missile technology before moving to BT's 'Foresight and Futurology Unit', where, among other things, he invented text messaging. Since 2007, he's gone it alone, using his expertise to forecast the social, political and environmental impact of technology. On his website Pearson claims an 85 per cent success rate for 10–15-year predictions and boasts hundreds of inventions (or at least ideas he's independently come up with), from satnav and interactive TV to vibrating fabrics and something called 'active contact lenses'. With the advances in genetics, nanotechnology and robotics, Pearson suggests that we are on the cusp of some mind-boggling developments, tantalising his audiences with the likes of talking dogs, hamsters 100 times cleverer than Einstein and 'smart yogurt' with the same brainpower as the whole of Europe.

Like many futurists, though, most significantly Pearson believes that we are fast approaching the point where technology will fundamentally alter our species, giving us vastly superior capabilities. Already, we can access pretty much all recorded human knowledge via our smartphones. The same device helps us keep track of our social networks, lets us speak to people on the other side of the world and tells us how to get to where we want to go. And the technology is only getting more advanced.

Today our smartphones are more powerful than the computers that first put man on the moon. As Pearson says, it's a positive-feedback loop: as our technology gets smarter, it enables us to come up with something even smarter. He predicts that within a decade we'll have conscious robots that will become our friends and lovers. Not only will they be capable of independent thought, but they'll be able to upgrade themselves, triggering a runaway process of self-improvement. Moreover, using microprocessors and plugging ourselves directly into the internet, we'll increasingly integrate with technology, becoming superhuman. It's what's known in the business as the point of singularity – when man and machine become one. Best of all – depending on your view – Pearson is confident that anyone younger than forty will live forever. Although he's also calculated that we'll most likely be extinct by 2067. All in all, with his diamond-shaped glasses, here was a man who could tell me where we're heading.

It turned out that Pearson didn't live in an oxygen tank high up in a skyscraper, as I'd imagined, but in a mock-Tudor suburb somewhere on the edge of Ipswich. There wasn't even a doorbell.

'Andy Scott,' he said, as he opened the door. He'd traded his diamond glasses for round ones but, metallic red, they still gave him an edge. As he led me through to the kitchen, I was on the lookout for gadgets and buttons, maybe a talking dog, but there was nothing – not even a voice-activated kettle. Pearson explained that he came originally from Northern Ireland; his work with BT had brought him to Ipswich and he'd ended up liking it there. We sat at the kitchen table, overlooking the bird-feeders, as he recalled how it was watching the first moon landing in 1969 that had sparked his interest in the future. Now he uses his background in engineering and systems analysis (piecing together complex processes) to work out what's coming.

I asked when the future is, what sort of timeframe we should be thinking about. 'Generally, the future's around thirty years away,' he said, explaining that this is usually the time it takes to get an idea from conception to wide availability. It's perhaps no coincidence, then, that *Back to the Future Part II* looked ahead thirty years, to the unimaginable world of 2015, though the standout innovation was the fluorescent hoverboard. For his part, Pearson said that there's little in today's world that he hadn't seen coming back when he started out at BT in the early 1990s. Even though the internet was only just becoming publicly available, he'd predicted that we'd end up doing most of our business and interactions online.

'Is there anything that's surprised you?' I asked.

He thought for a moment, perhaps not wishing to dent his record. 'I suppose we didn't expect the degree to which people are willing to give up their privacy,' he said. 'I think some of it is that we don't always realise.' (Boiling the frog.) He suggested that it was a generational thing, with sixteen-to-twenty-five-year-olds not having the same concerns. 'Now, if you don't have drunken photos everywhere, people wonder what's wrong with you. It's a sort of reverse stigma.'

This raises a wider point about how social media is changing the way we interact. Prompted by the likes of Facebook and Twitter, we now freely disclose what's going on in our minds and lives to ever-expanding networks. We no longer wait for people to ask 'How are you?' As the founder of Facebook, Mark Zuckerberg, says: 'The question isn't "What do we want to know about people?" It's "What do people want to tell about themselves?"' But beyond concerns about privacy, the issue is that we've become accustomed to disclosing ourselves in a particular way. As Greenfield has observed, reflecting the ideas of Erving Goffman discussed in Chapter 4, social media is very much life on the 'front stage', so we inevitably put our best foot forward, airbrushing our lives. When Facebook asks what's on our mind, we generally stick to the positive, posting pictures of us holidaying, partying or enjoying a fancy meal, rarely ones of us sitting alone eating tinned soup. And so we end up doing what Julian Barnes warned against: judging the inside of our own lives by the outside of others, scrolling down all of the achievements and parties, wondering what's gone wrong. As we live more of our lives online, social psychologists have been led to worry that we'll become conditioned to sharing only the positive, measuring our own lives accordingly, and less inclined to say what's going on on Goffman's 'back stage'.

For his part, Pearson was equally baffled about what people feel compelled to share online. His big concern, though, was that social media was stifling debate. Rather than flattening barriers, it can create invisible ones. When it comes to life online, we choose whom we interact with, silently deleting anyone who annoys us. As Pearson put it, we're turning into a 'bathtub society', divided between left and right.

'Why can't people disagree any more?' he said.

As a futurist, Pearson believes that it is his job to stay neutral. But I sensed that his profession made him something of an outsider or at least took him down paths that others couldn't easily relate to. That evening, he was off to the British Academy to give a talk on sex with robots.

'I don't see a problem with it,' he said.

Sooner or later – probably within the next decade – Pearson predicts that they'll have their own minds and personalities, that we'll fall in love and marry them. I was sceptical, but he reminded me of *Star Wars* and how we get emotionally attached

to the likes of R2-D2. Apparently, a psychologist was also attending the event to speak against the idea.

'But she's missing the point. The big thing is that, by 2050, *you*
could be the robot,' he said, eyes widening. 'Your brains' – that
is, my brain and that of the robot – 'would be networked so you
could inhabit each other's mind and feel what they're feeling.'

I was struggling to get my head around all this (would we
feel both sets of sensations?), but also concerned that we were
jumping ahead. Before getting on to changes in the brain and us
becoming robots, I wanted to find out more about the different
technologies and devices that will impact the way we interact
and greet each other. I mentioned that I'd been reading about
various developments designed to enrich our remote communications, such as 3D video and holograms. Using ultrasonic
radiation, for example, a team of scientists from the School of
Frontier Sciences at the University of Tokyo have taken holo-
graphy to a new level, creating a hologram you can both see and
touch, with guinea pigs reporting a bubble-like sensation at the
fingertips. In the long term, their vision is to make it possible
to shake someone's hand from thousands of miles away. And,
recognising the importance of touch to our communication,
a fashion company from London has come up with the 'Hug
Shirt', which allows wearers to send and receive hugs via a
Bluetooth connection. The shirts are embedded with sensors
that can detect the strength and duration of a hug, even the skin
temperature, though the jury is still out as to whether they create
the same psychological effects as the real thing.

Pearson explained how during his time at BT he'd toyed with
the idea of 'emotional clothing' that could be activated by text
messaging. But this stuff was already behind the times. The
real potential came with 'active skin', which will give us a whole

range of capabilities. The technology hasn't been proven yet, but Pearson suggested a couple of possibilities: either touch-sensitive electronics printed onto our bodies or being injected with a gel that contains millions of tiny computers, which will sit side by side with skin cells. Originally, he'd thought of this in terms of health benefits, such as our being able to monitor our blood chemistry, but he realised that it could also transform our remote communications, giving us full tactile stimuli. However, he wasn't as keen on this.

'Personally, I'm a big fan of email and the phone. I don't feel the need to see or touch people all the time. I'd find that sort of contact over the top. Like lots of people in IT, I'm slightly Aspergic.' I wondered if that partly explained why the likes of the Hug Shirt and emotional technology are yet to get off the ground.

I asked Pearson about his 'active contact lens' idea. His website includes the original design he came up with in 1991, involving a series of lasers and micromirrors that project an image onto the retina. The basic idea is that it will display any kind of visual information right in front of you, whether an app or an entire virtual world. 'You'll be able to walk down streets paved with gold,' he smiled. But, as he explained, we'd also be able to see people as we want, changing our partners into our favourite star or whatever takes our fancy. Moreover, we'd be able to talk to people across the world as if they're sitting next to us. They'd even work with our eyes closed, so we really would be 'always on, always connected'.

When it comes to meetings and greetings, though, Pearson said that the best thing he could offer was what he called the 'ego badge', which he came up with in 1995. Consisting of computer chips and a transmitter, it would be a wearable device that would broadcast our personal information – the equivalent of a

Facebook profile – to everyone around. So whether at a party or walking down the street, the badges would interact, alerting us when they've found a match, whether a business client or date. They'd even work out a time and place to meet.

'You wouldn't need to worry about being shy or waste all that time with small talk,' Pearson said. He mentioned how he spent a lot of time at conferences where there were special networking sessions. I could feel his pain. 'You can end up talking to the worst bore in the country,' he said. The personality badge would cut out all that. 'You'd have already established a common bond and friendship. It would be an electronically enhanced greeting.'

I could definitely see the appeal. But I also wondered if it would turn us into even more of a bathtub society, only ever interacting with people who share the same mindset.

'If the technology is there, why don't they exist?' I asked.

Pearson thought for a moment. 'I've no idea.' But he then added that they sort of do, just in a different form. Our smart-phones, social media and dating websites already do something similar. The ego badge would just keep us on high alert.

'What about mind-reading?' I asked, thinking that this was the Holy Grail. Going back to the joke at the beginning, we'd no longer bother with all our empty well-being enquires. There'd be no place for doublespeak, false modesty or deception. We'd no longer struggle to read people's expressions and piece together the clues. The 'theory of mind' would be a reality.

As Pearson explained, the technology is some way off, but we're getting there. Through attaching electrodes to the scalp, scientists can already read and process neuron activity via a computer. There are now robotic arms and wheelchairs that can be operated through thought alone. And, recently, scientists crossed the Rubicon, connecting two brains via a computer.

In 2013, a team from the University of Washington in Seattle created a brain-to-brain interface that allowed two people to play a computer game cooperatively: one person thought about the hand movements for shooting a rocket, while the other person received a magnetic pulse, causing an involuntary twitch in their wrist over the touchpad. Many futurists believe that by 2045 we'll be able to read and experience each other's emotions, even trading them online. So you'll be able to 'feel' what it's like to be at Glastonbury, even what it's like to perform at Glastonbury or whatever else you're up for. And you'll know whether someone really is pleased to see you.

Pearson is looking forward to direct thought transmission, though he doesn't think it will be for everyone. 'Imagine people with Trump-sized egos who can't wait for people to see and admire what's going on in their brains.'

Inevitably, all of this talk of mind-reading brought us back to some of the bigger questions about how science is changing us and where our species is heading. Like many futurists, Pearson believes that we are on the cusp of the biggest change in human history. Genetics will allow us to customise our DNA, giving us designer babies with preselected looks, talents and intelligence. But the biggest boost will come as our biology merges with technology. We'll have bionic limbs and binocular vision. Most of all, we'll be able to enhance our brains, wiring them directly to computers, unleashing our intelligence to infinity. Pearson predicts that this will be the route to immortality. We'll simply upload our minds online, so that when our bodies pack up we'll be able to download them onto a robot and go on as normal.

'As I like to say, death will no longer be a career problem.'

'We'll redesign nature,' he said. As he explained, after however many billion years of genetic evolution, with all our improvements relying on random mutations, we'll take control of the process, directing it as we want. Or, put another way, having done so much to disprove it, science will finally give us intelligent design.

'We'll become *Homo optimus*. You'll be Andy Scott Mark II,' he said.

Yet, for all of his excited predictions, I wondered if Pearson really believed in any of it. No doubt there are those who do. During my research, I'd come across the Transhumanist movement, which is committed to using technology to better our species, whether through enhancing our intelligence or making us immortal.

'The big issue is human nature,' Pearson said.

'But I thought we'll be able to change human nature for the better,' I said.

'That's the crux of the problem,' he said, gripping the table. 'Who decides what's better and which direction we should go in?'

Pearson reasoned that human culture was like a random walk in the absence of a guide. We could never trust one generation to make the right decisions about our future direction.

'When I was at school, I was told that abortion and euthanasia were wrong. Now look at what people think. Our genes could end up being a matter of political fashion. I find that thought horrifying.'

Rather than dispelling my concerns, Pearson was helping to clarify them. Moreover, his own doubts were compounded by the current state of political leadership. 'I am an optimist, but I look around and ask whether we have the wisdom or knowledge to use all this technology responsibly. *Do we hell*. It's like giving a chemistry set to a three-year-old.'

Pearson was also uncomfortable with the idea that some of us will have superhuman capabilities, while others won't. 'There'll be some people with an IQ of a billion, who'll monopolise or want to wipe the others out.'

I could see how we'd end up with a whole new class system, only this one would have a material basis. It'd be like the knights and serfs all over again, with some of us scurrying around, bowing and kowtowing. Maybe the handshake isn't so safe, after all.

'We're on a very dangerous path,' Pearson concluded.

I was beginning to wonder who the real technophobe was.

Pearson gave a wry smile. 'As a technophile, I am excited about the future and all the new technology. But as a human I'm disturbed by it.' With his red glasses, suddenly it was like he was a flashing warning light.

'What about living forever?' I asked, hoping to end on a positive note. 'You said in a talk that you're really annoyed about being part of the last generation to die.'

'I don't want to live forever,' he said, looking at me like it was a strange question. 'I'm in no hurry to die, but I'm quite happy that my life will end at some point.' As for electronic immortality (living online or inside a robot), Pearson conceded that it won't be the same.

'In any case, there are disadvantages with immortality,' he said, explaining that we'd overload the environment and clog up culture. He also suggested that we never really become free until our parents die, since we're always looking for their approval. So we'd end up depriving the next generation of the same freedom. 'It's our duty to bugger off and die at some point.'

———

For all that technology will shape our behaviour, the bigger point to emerge from my conversation with Pearson was how it will (or at least should) always reflect deeper social and political trends. And given what we know about the greetings of the past, it's these that will ultimately determine how we greet in the future. The problem is that this sort of stuff is notoriously hard to predict, perhaps harder than ever. If I'd bet a pound that the UK would vote to leave the EU, Trump would be elected and Leicester City would win the Premier League, I'd have won £4.5 million, but I didn't and nor did anyone else. But another way at the future is to look back, identify trends and project forward. When pushed, Keith Thomas offered a few thoughts. Picking up from our discussion on greetings inflation, he could see how the cult of informality and friendliness would continue. Gradually, we'll lose the older and transitional generations that grew up accustomed to the standards of restraint and reserve. I even suggested that one day it would be common for Englishmen to greet each other with kisses again. Thomas also predicted that the English language will make further inroads, so sooner or later we'll have truly universal greetings.

None of this is surprising, but it does get at two longer-term trends that have deep undercurrents and have done much to shape our interactions: the decline of hierarchy and the growing homogenisation of our cultures. Even if by some measures inequality is growing, the Enlightenment ideals of freedom and equality have been on the march, with some arguing that they are fundamental to the human condition. As the Cold War came to an end, the American political scientist Francis Fukuyama even talked of the 'end of history'. It wasn't that things would stop happening, he suggested, but that the ideological wars were over: liberal democracy had won, proving itself to be the final stage of our cultural evolution, best satisfying the human condition.[3] Put differently, this was the future with a direction. By this logic, we'll keep shedding our formal hierarchies and class systems. At the everyday level, we'll increasingly be on first-name terms and shaking hands. And there'll be no need for symbols of counterculture, since there'll be nothing to rebel against.

In some ways, this can be seen as the latest – or final – incarnation of the trend towards homogenisation. As a number of historians who've pieced our story together have observed, since the first agricultural revolution the world has gradually converged around fewer and fewer norms. Whether through cooperation, coercion or colonisation, the big fish have tended to swallow up the smaller fish. If we use language as a proxy for cultural diversity, the pattern is clear. With the rise and fall of various empires, a few imperial languages, such as Arabic, Spanish and, increasingly, English, have come to dominate. Between 1970 and 2005 alone, the number of living languages fell by 20 per cent.[4] Today, it's estimated that one language dies out every couple of weeks or so. From this, we can deduce that other aspects of culture are disappearing at a similar rate. When I met Adam Kendon, for example, who'd spent time working with the Walbiri tribe in Australia, he described how the people of this once-unique culture now mostly wore jeans and T-shirts and are more likely to greet each other with a handshake than a genital grab. It's a familiar pattern across the world. You only have to look at somewhere like Ulaanbaatar, the capital of Mongolia, and compare how it is now to what it was like twenty years ago to see the direction of travel.

Taking the long view – and putting aside all the bloody history – the idea of us converging around one culture has a pleasing logic. It's humanity come full circle. Having started out as a small population in East Africa, we're finally regrouping as one interconnected people. Only 200,000 years later, there are now

7.4 billion of us (11.2 billion by 2100) and the whole world is our home. It's what struck me when I visited the British Museum, as I wandered between the artefacts, which tell the story of our different cultures and civilisations. As I reached statue overload, I found myself looking at the crowds of people. They were from all over the world, but through squinted eyes blurred into one: dressed in jeans and a shiny puffer jacket (it was November) and clutching a smartphone. The trend might be uneven, but as the Black Eyed Peas song puts it, we're becoming 'one tribe'.

For many, there are aspects to all this that are a cause of concern and regret. We're losing cultures that have grown over thousands of years. But, more than that, we worry about what kind of culture we're left with, whether it's culture*less*. First and foremost, we're all consumers now, with a few big brands symbolising our global identity. 'Hi, how are you?' can feel less like recognition of us as fellow human beings and more like a prod to buy something.

But while we might bemoan the blending (or blanding) of our cultures, in some respects it merely reflects that most human impulse: to belong. This, after all, has largely been what culture is about. Inevitably, then, as different peoples connect, they converge, eventually losing some of their distinctive habits and customs. Moreover, going back to the microsociologists, having a common set of everyday behaviours that we recognise and trust – or etiquette – can be seen as the starting point of a functioning society. So it's no surprise that for our McWorld we should have a McGreeting: the handshake. And the fact that it's English that has become so universal is beside the point. It's no longer about empire, but the fact that the different peoples of the world have become ever more connected and interdependent.

All that said, who today would dare predict that this 'one tribe' vision will be the story of tomorrow? Soon after Fukuyama pronounced the end of history, another American political scientist and his former mentor, Samuel Huntington, published his own look at the future, one that had a very different ring to it: *The Clash of Civilizations*. In bringing us closer together, Huntington argued that globalisation had made us more aware of our differences and would divide us along cultural lines: what he foresaw as a clash between the West and the rest.[5] While Huntington was criticised for overlooking our increasing interdependence and risking a self-fulfilling prophecy, against the rise of various extremisms and autocrats, the world can look increasingly fractured. Moreover, thinking about our own bathtub society at home, Pearson foresaw the coming of what he called the 'Great Western War'. Even Fukuyama worried that liberal democracy faced dangers, not just from autocracy and fundamentalism, but also from within. As we abandon all of our irrational beliefs and respect each other's rights, we'll end up self-absorbed and bored, lacking a sense of belonging and common purpose, reverting to old ways. He reminds us of the way that many people reacted when the First World War broke out: they took to the streets and celebrated, exercised by a feeling of solidarity – and this at a time of relative prosperity and optimism.[6] So, while we can't easily prophesy, perhaps the biggest lesson of history is that we should never get complacent.

———

If the message of our recent past is cautionary, another way at the future is to look even further back. This at least is the approach of evolutionary psychologists. The basic idea is that since *Homo sapiens* first wandered the African plains 200,000 years ago, however much culture we've accumulated, physiologically we've hardly changed. We may have grown taller, but our brains are pretty much the same. If you were to pluck a baby from a prehistoric cave and bring them up in a twenty-first-century city, they'd grow up with the same hopes and anxieties, just as glued to their smartphones as the rest of us. More to the point, having spent the vast majority of our time living as hunter-gatherers, roaming around in small bands and tribes, this is what our brains – and all the capabilities and emotions that go with them – are adapted for. The 12,000 years since the first agricultural revolution just hasn't been long enough for evolution to catch up. All of which means that we've been left with traits that are unsuited to modern living. So, as if still searching for our next meal, we have the tendency to gorge on high-energy foods; we're constantly seeking out ways to satisfy our reward system, so end up taking drugs and jumping out of planes; and we're still primed for savannah living where danger was never far away, so get overly anxious and fear strangers.

Looking to the future, the implications of this are clear: if we're stuck with our ancient brains, in order to live healthier, happier and more fulfilled lives, we should reconnect with or at least respect our hunter-gatherer selves. It's an approach that many of our lifestyle gurus are taking. Nutritionists now advise us that the best diet doesn't come from a constant supply of ready-made meals, but is a mix of nuts, seeds, leaves and berries, with the odd bit of meat, and days of eating less, just like our foraging ancestors. And, after years of being sold bouncy trainers that encouraged us to run on our heels, we're now being told to run on our toes, as if running barefoot across the savannah plains (so we end up buying £100 plimsolls instead). We're also slowly realising that our minds and bodies aren't best suited to sitting in front of a computer all day, after all.

Most of all, though, psychologists tell us that our well-being depends on keeping social, making sure we have friends and family around us, so we feel part of our band or tribe. It's why, despite being surrounded by millions of people, city living can feel lonely: our brains are geared up for social networks of 150. So we set up clubs and societies or take up new hobbies. It doesn't really matter what the activity is, so long as we're part of a group. It partly explains all our countercultures and subcultures – really they're just microcultures, connecting people at a manageable level. In effect, we're endlessly recreating those social units that our brains are wired for. Moreover, as Robin Dunbar suggests, within our social networks we'll always have layers of friends, with a small inner circle, probably never more than five – it's how our brains have always worked.

And so it is with our greetings. Fundamentally, they will (should?) remain the same. For all our cultural invention, at core they'll always be about affirming and testing our bonds. We may no longer be roaming the open plains so much or surrounded by danger, but we're still wary of strangers, and those first moments of interaction can be emotionally charged. So even though we might not be carrying spears or swords, our physical gestures can still help to pacify the situation. And we'll forever save our most intimate greetings for the people we're closest to and have special greetings among our groups, while reverting to more iconic rituals with those outside – just as we did during our hunter-gatherer days. Moreover, our greetings will continue to express our hierarchies and social status. However much we've stripped back our formal structures and champion meritocracy, we'll still have our pecking orders. Whether it's our parents, our boss or the most popular person at school, there'll be people we pay special deference to. We may not bow as much or as deeply, but we'll still find ourselves performing little acts of submission, whether a nod or, conversely, standing up when someone enters the room, making a display of inconvenience.

Looking into our evolutionary past might also help to explain our greetings inflation. Having shed our formal empires and some of our prejudices, we're simply going back to how we used to greet, using more physical and intimate gestures. It's perhaps why Englishmen have started to hug and may end up kissing again. Most of all, though, looking at our closest relatives and all our own greetings, we'll continue to revert to the various infant-ilisms that characterise the closest bond of all, between mother and child. All in all, when it comes to predicting the greetings of tomorrow, as with so many aspects of our behaviour, it really will be a case of back to the future.

Goodbye!

'm standing outside the Embankment Tube station down by the Thames. It's a cold January afternoon, not long after Christmas, almost exactly a year after my visit to Heathrow, where my research began. The book is now late, but there's one final task I've set myself, which I've been putting off: saying hello to everyone I walk past. I found myself committing to the idea when reading *When Strangers Meet* by the American social commentator Kio Stark. Living in New York, Stark has become an evangelist for talking to strangers and the power of small interactions. She encourages us to take risks and stretch out of our comfort zone, suggesting a number of 'expeditions' to get us going, of which this is one. Her instructions are simple: find a reasonably populous place such as a park or city sidewalk and say hello to everyone – 'All of them.' The Embankment seemed like the perfect spot: there's a long, wide path and plenty of people, but not too many; and it had personal significance, since it's where I'd had that painful encounter with my sister's friend all those years ago. In a way, I felt like I'd come back to exorcise my demons. The problem was that I was dreading the idea. I've bungee-jumped, run marathons and chased a burglar through London with a squash racquet, but this was making me feel uniquely odd. I sat on a bench overlooking the river and watched the people walking by. What could be so bad about saying hello to a few strangers? *'Come on, you wimp...'*

—

And so we've come to the end of our journey. It's been a bit longer than I'd planned, so the big question is: what have we learned? Starting out, I had some basic questions about greetings, what they are and why we do them. We've come a long way since I was standing in Heathrow, but there's a simple answer: it depends on who you talk to. It's a bit of a cop-out, but having met various experts, the overriding if obvious point is that the way we understand the world and human behaviour is shaped by what we do and the subjects we study. So going back to the etiquette and body-language experts, our greetings are about creating the right impression, whether to demonstrate good manners, show our class or get ahead in the world. For microsociologists seeking to understand what holds society together, greetings are a matter of managing our interactions. More fundamentally, we use these little rituals to acknowledge each other as social entities. For social psychologists, greetings are the moments our minds become socially engaged; how we do them is shaped by what we think others are thinking, so they become mini performances. If we take an evolutionary perspective and look to the rest of the animal world, we find similarities that suggest a common origin. Whether comparing ourselves to chimpanzees or to pigeons, it's clear that many of our rituals recall the closest relationship of all, that between mother and child. We continue to use them to reaffirm and test our bonds, to signal our status and show that we want to get along. For neuroscientists, the important thing is how all of this registers in our brains as feelings and sensations. To explain all of our differences, anthropologists interpret our greetings against our cultures as a whole, while, looking back, historians show how they reflect the social and political make-up of a particular time.

At times, all these disciplines can end up talking past, even against each other. But without too much bodging, it seems clear that greetings are all of the things above. These first moments of interaction are packed with significance, revealing not just the nature of our relationships, but the societies and times we live in. Perhaps the real value of this book, then, has been in taking a small slice of human behaviour, looking at it from all angles, and showing that the various disciplines all contribute to our understanding, complementing rather than cancelling each other out. As I've gone around meeting the experts, delving into their different areas, I've been struck by how I've never come away thinking that what I've been told sounds contradictory or wrong, whether coming from an anthropologist or a zoologist. Stepping back, it's reaffirmed my view that it's only by looking across the fields that we can gain a full picture. And what's true for greetings must be true of all aspects of human behaviour, from making art to fighting wars.

This brings us back to an aspect of greetings that had originally sparked my interest: the fact that they are both universal and particular, something we all do but each in our own way, whether as individuals or as cultures. I now realise that this gets at the most fundamental question, one which has ripped apart academic communities: the extent to which human behaviour is innate or learned, a matter of instinct or of culture, nature or nurture. In a small way, looking at greetings has shed some light on this question. Perhaps reflecting the times, but also my own experience of the highs and lows of life, I'd started out thinking that there was a common-sense view that recognised the power of both. Now, though, I don't think this goes far enough. I'm not sure it's helpful or even accurate to draw these kinds of distinctions. To suggest that humans have no nature or at some point transitioned into an age of culture is deeply distorting and our greetings partly prove the point. As we've seen, taking the approach of evolutionary psychology, the fact that greetings are universal suggests that they are an adaptation, helping us to survive and reproduce. With the help of various ethologists, we've seen how they have functioned as signals of cooperation – which is fundamental both to our survival and our reproduction. But even if you don't like this logic, if we look across the animal world, particularly at our closest relatives, it's obvious that many of our rituals can be traced deep into our evolutionary past. And if you're in doubt about our connections to the animal world, go on YouTube and watch the documentary about the group of chimpanzees at Arnhem Zoo.

The fact that our greetings are so particular comes down to our giant (relatively speaking) and malleable brains. Above all, it's our grey matter that has defined our species, giving us the mental capabilities to live in bigger groups, learn and pass on new behaviours and, as a result, adapt to and thrive in new environments. This is what gives us our different ways of living (cultures) – yet there is no dramatic break with nature here. As a result – maybe even just a by-product – of all this, we've ended up with a higher level of consciousness that enables us to reflect on our own thoughts and question what life is all about. This is the bit that has generated so much knowledge and creativity, as well as nonsense. It's also the bit that can get us into a muddle, such as not knowing what to say or whether to go for one kiss or two. Yet for all that our greetings vary from culture to culture and are shaped by the social norms we grow up with, they've retained similarities with the rest of the animal world, both in form and function. And for all that the power of reflective thought has enabled us to expand and hone our behaviour, it seems that we are still saddled with the same brains we had 200,000 years ago. We may be more broad-minded and worldly than ever, but we still make snap judgements based on people's appearance. We're still biased towards members of our own group and prone to prejudice towards outsiders. This isn't to say that any of this is right or even much use in today's world, but we should at least be aware of it. After all, no one protests when commentators talk about our 'social instinct' or 'essential need' to belong. When it comes down to it, however much anthropologists help us appreciate our diversity, they tend to focus on the elements of the human experience that transcend our differences, such as reproduction, death, kinship and marriage. In the end, when we look at people as individuals, fundamentally we are all the same – unless, of course, the Transhumanists get their way.

All of which sort of brings me back to the other inspiration behind this book: awkwardness. In many ways, researching and writing this has been cathartic. More than ever, I've come to realise that I am not alone in finding greetings fraught with confusion and embarrassment, even to the point where they can put us off our interactions altogether. And more than ever, I blame my Englishness. It was brought home when, towards the end of my research, I visited the Japan Society in London. Having discussed Japanese greetings and how they rarely involve any kind of physical contact, I asked my host how she found the English, suggesting that we share some similarities. She agreed, but also said that she found our greetings particularly difficult, since she never knew what to do, despite having lived here for years. By contrast, she'd also lived in Colombia, where people always kiss, which was much more alien to her, but she actually became comfortable with it. She knew what to expect.

Coming back to the title of the book, it's clear that the kiss causes most problems. I'd originally assumed that this was a recent dilemma, but taking the longer view, it seems that it's often been a live issue. Back in the seventeenth century, the *Spectator*'s agony uncle fielded questions about the 'parameters of lawful kissing', while some men were said to have been driven to violence or madness by the sight of their wives kissing every 'handsome fellow' who came to their house.¹ Getting to the nub of the matter, when a group of historians, including Keith Thomas, met in 2000 to discuss the kiss in history, they kept coming back to two themes: ambiguity and transgression.² The basic issue, distinguishing the kiss from other gestures, is that it can mean so many different things – tenderness, deference, love – and so is ripe for misunderstanding. Even though the historians found that the kiss's meaning has changed throughout history as attitudes towards the body have altered, as Thomas concluded, in ancient times as now, the lips were an erogenous zone, whatever our cultural conventions.³ The interesting thing, then, as Raymond Firth puts it, is 'the way in which Western societies are prepared to skate on thin ice in juxtaposing so closely an erotic and merely a social gesture, while most other societies have avoided this problem'.⁴ For Thomas, all this explains the double kiss: it helps to draw a line; it's as if the follow-up kisses cancel the meaning of the first. I could see the logic, but still couldn't shift the feeling that it's a bit pretentious.

There's no doubt that we English have special issues. And, sadly, as I've come to understand them better, I can't say I've become any less awkward. However much we might be world leaders, though, it's clear that we're not alone. In fact, I've come to think that awkwardness and embarrassment distinguish our species. Much of it stems from our unique powers of self-consciousness – our lives are full of decisions and worrying what others think of us. But while all this can be debilitating, it can also be a good thing (to a degree). Our awkwardness and all the physical side effects demonstrate that we understand that there are certain social norms and personal expectations, even if we're not sure what they are. For all our uncertainty, we're at least proving that we want to be accepted and trusted. So, at times, the best way to tackle awkwardness might be to embrace it.

This brings me onto the advice bit of all this. I'd set out hoping to solve our uncertainties and cure the world of greetings anxiety, even to find the perfect greeting. I soon realised that this was never going to happen. For all that we have standard rituals, our greetings are just too complex, reflecting the nature of our relationships, the culture we're from and the times we find ourselves in. And that's without taking into account us as individuals, our personalities and changing moods. In the end, our etiquette guides mostly capture what we already do and are trying to keep up, when they're not obsessing about class. And I'm still wary of the body-language gurus and idea that we can shape what people think of us with a few easy tweaks. We're always on the lookout for phoniness, suspicious that we're being manipulated. After all, it's why the etiquette and body-language experts go on so much about sincerity and authenticity. That said, I realise that I can't get away with writing a book about greetings without offering some sort of guidance – even if, after all this, I'm not sure I'm any better at them myself. Stepping back and skirting the details, there are six basic points:

1. *Greetings are important.* Hopefully, this will be obvious by now. There's no need to rehearse all of the arguments, save one: our greetings acknowledge other people as social

entities. It's enough to explain why a snub can hurt so much and why we neglect them at our peril.

2. *Be wary of your judgements.* We all like to think that we're great judges, but in truth we're also prone to being guided by irrelevant information and irrational biases. The jury might be out on the strength of the handshake, but we should at least realise that there's nothing universal about it. We need to recognise that our brains are susceptible to mental short cuts before giving in to them. Though there may be no harm in strengthening your grip if it feels like, in Emily's Post's words, a 'spray of seaweed'.

3. *Don't try too hard.* While it's important to have an under-standing of the different customs of different cultures – mostly to help keep our judgements in check – if we try too hard to fit in, we risk looking like an imposter. We might appreciate it when people make an effort to embrace our culture, but we also recognise that it can take a lifetime to learn all of the quirks and absurdities that bind us. So while it's often best to go with the flow, there's no point pretending that we're something we're not.

4. *Keep it open.* There are few aspects to greetings or body language that are universal, but as a rule we are looking for signals we can trust. So it's a good idea to make our approach with an open posture and intentions clear (which is easier said than done, if we're full of doubt). To reduce the uncertainty, it can be helpful to prepare the way with an outstretched hand if we're going for a handshake or open arms for a hug, so we don't spring any sudden surprises.

5. *Reciprocation rules.* As with most things in life, we tend to notice when there is a mismatch in our greetings. We feel annoyed if someone doesn't say hello back or return our

well-being enquiry, and are conscious when our physical rituals get out of sync, whether it's the number of kisses or the strength of a handshake. All this equalising helps to build a rapport, but it also gets at something more fundamental: our sense of fairness. If we don't reciprocate in kind, it suggests either a breakdown of etiquette or difference in status.

6. *Don't worry if it goes wrong.* This may sound like empty reassurance, but when it comes to our social blunders, we should always keep in mind 'protagonist disease' and the 'spotlight effect' – people are less focused on us than we think. And remember that a bit of awkwardness is a good thing, or at least isn't always perceived in a negative way. It can be better to announce our uncertainty and mistakes than to try to cover them up.

And now the phrase 'practise what you preach' comes to mind.

So, if that's greetings, what about goodbyes? People have often asked me, jokingly, if there will be a sequel. It's something Adam Kendon brought up as he waved me off: 'You know, you'll also have to look at goodbyes. They're very interesting.'

'The mirror of greetings,' I found myself shouting back.

'That's it,' he said, turning away.

So, to my relief, no need for another book. However you interpret them, goodbyes are basically the mirror image of greetings – the other half of the 'ritual brackets' that frame our encounters, as Goffman put it. While we use greetings to gain access to each other, we use goodbyes to escape. But as with greetings, they are much more than that. They signal that our encounters have been acceptable and we vary them according to the nature of our relationship and when we will next see each other. In parting, we revert to small talk (weather, road directions and the like), say how we'll be in touch soon, even if we have no intention of doing so. The happiness of meeting is matched by the sadness of separation, which is graded according to when we'll see each other again. Interestingly, while our greetings reveal our connections to the rest of the animal world, our goodbyes suggest a difference. Speaking to Goodall, she explained how the odd thing was that, when it comes to parting, chimpanzees just get up and leave without any fuss. The suggestion is that this demonstrates that only humans can think ahead and contemplate a time of separation. Then again, in his most recent book, Frans de Waal describes how one of the chimpanzees he got to know would go around her group giving kisses before leaving for the day.[5] And so the debate goes on.

———

Finally, back to the Embankment. I sat on the bench for a few minutes, staring at the London Eye, wondering why I was dreading such a harmless thing. Though there were no signs warning me against saying hello to strangers, the obvious point was that I was about to break an unwritten rule of city life. Generally, it's only drunks and people with clipboards who go around saying hello to everyone, especially in London. So I'd be exposing myself as someone who doesn't know or respect the norms, as a sort of social deviant. But even though I felt like I'd got to the root of the problem, just like when I found myself standing on the edge of a bridge high above the Zambezi River with a bungee cord around my ankles, there was no backing out. I stood up, took a breath and told myself that I'd start at the next traffic lights. A man came towards me who looked unthreatening enough. As

we came within a few metres, I looked up, made eye contact...
and immediately looked away. *Bugger.* A woman walked towards
me pushing a pram. *OK, this is it.* I looked up and managed half
a smile, but again pulled away, feeling like some sort of creep.
I tried again, but kept wimping out, averting my gaze as normal
or managing a hello under my breath at most. At one point, a
couple stopped me to ask for directions. It felt like I was being
mocked. *I'm the one who's meant to do that.* But I made a point
of asking where they were from (Brazil) and wishing them a
good trip, proving to myself that I wasn't completely incapable.
It didn't count, though. I made it all the way to the Millennium
Bridge without saying hello to anyone. I just couldn't do it. Even
though I'd failed, I told myself that it was a valuable insight into
the power of social norms. I could almost feel them pressing
down on my brain. It also made me realise how subtle and particular they are, since back in the small village where I'm from,
I have no problem saying hello to strangers.

But this was no way to end my book. So I turned around and
started again. I practised smiling and saying hello at a proper
volume, before choosing another set of traffic lights. First up
was a family. I cleared my throat, looked up at the dad, smiled
and said hello. He smiled back, clearly puzzled. But I'd done it.
And the funny thing was, I felt really good, elated even. A shorter
man walked towards me, with a purposeful stride. 'Hello,' I said.
This time he said hello back, as if there was nothing odd at all.
I found myself laughing as I walked on, thinking, 'Wow – that
was a really good one.' Next came a couple of teenage girls. They
also said hello back, though they looked marginally freaked
and I could hear them giggling as I carried on. I kept on going,
saying hello to everyone, even people with headphones. Some
just looked away, but it didn't matter. The good ones kept me

going. I could feel my stride lighten, as though the weight of social norms had lifted. I guessed that the dopamine and oxytocin were also surging in my brain, rewarding my efforts. It was becoming addictive. But, as Stark sets out, the next level is to make some small talk. As she suggests, it helps to find what's known as a point of triangulation: a third element, something to talk about that closes the triangle.[6] It's why dogs, babies and the weather can be so useful. I could feel my English sense of negative politeness and fear of intrusion pulling me down again, but I came to a couple of guys setting up a camera. 'Nice view,' I said. They looked up, a bit startled, but managed a smile and I carried on. I'd done it.

Now I just had to decide when to stop. For all that I'd dreaded the idea, a bit of me wanted to keep going, to feel liberated and experience the little high each time someone said hello back – that moment of togetherness, as Stark puts it. But I also wanted to meet my friend for a pint and to get on with life as normal. It's a bit sad really, but I won't pretend that I now go around saying hello to strangers all the time. Though I do think we fear it more than we need to. As Stark says, by saying hello, we pierce the anonymity, turning strangers into humans. Moreover, strangers or not, it's through our greetings that we open ourselves to each other – and that's no small thing.

Acknowledgements

When I first came up with the idea for this book, I had little idea of what it would involve or where it would take me. As it turns out, mostly sitting in my old bedroom in a small village in Suffolk, worrying if I'd ever finish. But the fact that I have and still seem intact comes down to the support of many people.

First, thanks to Peter Tallack of the Science Factory. Peter jumped in early with an email that I will never forget, expressing his enthusiasm for the project. Since then, he has guided and shaped the book, mostly by pressing delete, but also through his insight and encouragement. Thanks also to Tisse Takagi of the Science Factory for helping with the proposal and for some much-needed reassurance when my doubts were getting the better of me.

As I discovered, when it comes to turning an idea into a published book finding an agent is only half the battle. So I am very grateful to Nikki Griffiths, David Marshall, Peter Mayer and the rest of the team at Duckworth Overlook for taking a punt. In particular, I have been very lucky that Nikki's successor, Gesche Ipsen, has been so supportive and encouraging, guiding the book through to completion with great care and an expert eye. Thanks also to Alex Middleton, whose forensic copy-editing saved me from countless mistakes. And thanks to Tracy Carns, Chelsea Cutchens and Chris Cappello at The Overlook Press for helping to give my book a home across the Atlantic.

Despite all the hours and days sitting in my room, often with only a cat for company (and thanks to them), one of the best bits about writing this book has been meeting so many different people along the way. Thank you to the following for sharing their time and expertise: Simon Bearder, Robert Foley, Kate Fox, Chris Frith, Jane Goodall, William Hanson, Naoko Heckle, Akiho Horton, Uri Hertz, Adam Kendon, Ian Pearson, David Shankland and Keith Thomas. I have also plundered their work throughout my research, hopefully not misrepresenting it or them too much. Thanks also to Rebecca Perry, Chris Haslam, Rebecca Kent and Angela Matthews at Colchester Zoo for introducing me to their star animals. And thanks to all the people on the streets of London, or wherever else, who took time to talk to me and answer my strange questions. I am also grateful to my old supervisor, David Reynolds, for his continued support and encouragement. Even though he's had little involvement with this project, I've still benefited from the standards he set.

During the writing of this book (as with many other times), I have disappeared somewhat, so I am very lucky to have so many good friends with whom absence isn't necessarily an issue. Some have contributed directly to this book, whether knowingly or not; others have been full of encouragement; but most have just listened to my various anxieties and kept me thinking about other things in life.

Finally, thanks to my family. To my brother-in-law Justin for Friday pints and for once again reading my ramblings with a sharp eye. To my niece Lara for her wise words and encouragement. To my nephew Jack and other niece Katie (see!) for always humouring their wayward uncle and for keeping me up with the likes of the Dalham Olympics and *Bake Off*.

I have always resented being thought of as the spoilt younger brother, but in truth I have been incredibly spoiled by my siblings. My sister Lizzie and brothers Chris and Rob have been rocks throughout my life and never more so than during the writing of this book. There is no doubt that this project and the time off it's involved would have worried the hell out of my dad, but I can only think and hope that he is in here somewhere. As for my mum, it's difficult to put into words the combination of love, support, concern, frustration and impatience (though mostly patience) she has shown as I've returned home to write this book. Only to say that one of the big themes here has been how crucial the relationship between mother and child is – I'm just lucky (she less so) that I didn't need to read a book to know that.

Notes

1. It's a Minefield Out There

1 T. Geoghegan, 'Pecking Order', BBC News [website] (12 October 2007). Available at http://news.bbc.co.uk/1/hi/magazine/7040259.stm.
2 'Scientists Create Formula for Perfect Handshake', Newspress [website] (2010). Available at http://www.newspress.co.uk/public/ViewPressRelease.aspx?pr=23313.
3 The VeryBritishProblems (@SoVeryBritish) Twitter account is available at https://twitter.com/SoVeryBritish?ref_src=twsrc%5Egoogle%7Ctwcamp%5Eserp%7Ctwgr%5Eauthor. The Channel 4 series *Very British Problems* began in August 2015.
4 'Kiss of Death or Friendly Salute?', *Daily Telegraph* (14 August 2011). Available at http://www.telegraph.co.uk/women/sex/8699636/Kiss-of-death-or-friendly-salute.html.
5 Thanks to Aaka Pande!
6 S. Snow, 'Hug vs. Handshake: Navigating Awkward Salutations in the Workplace', Medium [website] (15 May 2013). Available at https://medium.com/@shanesnow/hug-vs-handshake-1c4f35dec45b.
7 J. Moran, *Queuing for Beginners: The Story of Daily Life from Breakfast to Bedtime* (London, 2008), p. 5.

2. What's in a Greeting?

1 A. de Botton, *A Week at the Airport: A Heathrow Diary* (London, 2009), p. 13.
2 See E. Goffman, *Behavior in Public Places: Notes on the Social Organization of Gatherings* (New York, 1963); *Relations in Public: Microstudies of the Public Order* (New York, 1971); *Interaction Ritual: Essays on Face-to-Face Behavior* (New York, 1972).
3 Goffman, *Behavior in Public Places*, p. 84.
4 E. Davey, 'A Bumpy Ride' (2015). Available at http://emdavey.com/2015/08/13/free-short-story/.
5 Goffman, *Interaction Ritual*, pp. 1–3.
6 Goffman, *Relations in Public*, p. 79.
7 R. Collins, *Interaction Ritual Chains* (Princeton, 2005), p. 47.
8 Ibid., p. 249.
9 T. Allert, *The Hitler Salute: On the Meaning of a Gesture* (New York, 2008), p. 21.
10 M. Mauss, *The Gift: The Form and Reason for Exchange in Archaic Societies* (London, 1954).
11 K. Fox et al., *The Language of Gift Exchange* [report published by the Social Issues Research Centre, December 2015].

12 M. Macmillan, *Seize the Hour: When Nixon Met Mao* (London, 2006), p. 33; Y. Xia and C. Tudda, 'Beijing, 1972', in K. Spohr and D. Reynolds (eds), *Transcending the Cold War: Summits, Statecraft, and the Dissolution of Bipolarity in Europe, 1970–1990* (Oxford, 2016), p. 45.

13 R. Firth, 'Verbal and Bodily Rituals of Greeting and Parting', in J. La Fontaine (ed.), *The Interpretation of Ritual: Essays in Honour of A. I. Richards* (London, 1972), p. 1.

14 A. Kendon and A. Ferber, 'A Description of Some Human Greetings', in A. Kendon, *Conducting Interaction: Patterns of Behavior in Focused Encounters* (Cambridge, 2009), pp. 153–208.

15 D. Morris, *Peoplewatching* (London, 2002), p. 112.

16 H. Kissinger, *The White House Years* (London, 1979), p. 1055; Macmillan, *Seize the Hour*, pp. 23–5; E. Ladley, *Balancing Act: How Nixon Went to China and Remained a Conservative* (New York, 2007), p. 207.

17 I. Eibl-Eibesfeldt, *Love and Hate: The Natural History of Behavior Patterns* (New York, 1971), p. 174.

18 Kendon and Ferber, 'A Description of Some Human Greetings', pp. 173–5.

19 H. Sacks, *Lectures on Conversation*, i (Oxford, 1992), pp. 4, 96.

20 Ibid., p. 552.

21 Kendon and Ferber, 'A Description of Some Human Greetings', p. 186.

22 Eibl-Eibesfeldt, *Love and Hate*, p. 170.

23 H. Ling Roth, 'On Salutations', *Journal of the Anthropological Institute of Great Britain and Northern Ireland*, 19 (1890), pp. 164–81.

24 *The Journals of Captain Cook*, ed. P. Edwards (1955–67; London, 1999), p. 52.

25 Roth, 'On Salutations', pp. 178–9.

26 Quoted in Collins, *Interaction Ritual Chains*, p. 79.

27 B. Bryson, *Notes from a Small Island* (London, 1995), p. 278.

28 J. Paxman, *The English: A Portrait of a People* (London, 1999), p. 126.

29 J. Moran, *Queuing for Beginners: The Story of Daily Life from Breakfast to Bedtime* (London, 2008), pp. 187–200.

30 K. Fox, *Watching the English: The Hidden Rules of English Behaviour* (London, 2004), p. 39.

31 H. Sacks, *Lectures on Conversation*, ii (Oxford, 1992), p. 205.

32 See 'Protocol Gift Unit', US Department of State [website]. Available at https://www.state.gov/s/cpr/c29447.htm.

3. How to Greet the Queen and the Elusive Art of Etiquette

1 C. Taggart, *How to Greet the Queen and Other Questions of Modern Etiquette* (London, 2014), p. 1.

2 J. Morgan, *Debrett's New Guide to Etiquette and Modern Manners: The Indispensable Handbook* (London, 1996).

3 Ibid.

4 W. Hanson, *The Bluffer's Guide to Etiquette* (London, 2014); P. Post et al., *Emily Post's Etiquette: Manners for a New World* (New York, 2011); J. Morgan, *Debrett's New Guide to Etiquette and Modern Manners* (London, 1999).

5 Hanson, *Bluffer's Guide to Etiquette*, p. 47.

6 A. Ross, 'Linguistic Class-Indicators in Present-Day English', *Neuphilologische Mitteilungen*, 55 (1954), pp. 113–49.

7 N. Mitford, 'The English Aristocracy', *Encounter* (September 1955), pp. 5–12.

8 E. Post, *Etiquette in Society, in Business, in Politics and at Home* (New York, 1922), p. 48.

9 Ibid., p. 48.

10 G. Devereux, *Etiquette for Men: A Book of Modern Manners and Customs* (London, 1902), p. 12.

11 F. Ings, *Etiquette in Everyday Life: A Complete Guide to Correct Conduct for All Occasions* (London, 1918), p. 23.

12 Ibid., p. 33.

13 Post, *Etiquette in Society*, p. 14.

14 E. Holt, *The Encyclopaedia of Etiquette* (New York, 1901).

15 Post, *Etiquette in Society*, p. 20.

16 Devereux, *Etiquette for Men*, p. 27.

17 Post, *Etiquette in Society*, pp. 19, 8.

18 R. Duffy, 'Manners and Morals' (introduction to Post, *Etiquette in Society*), pp. vii–viii.

19 N. Elias, *The History of Manners: The Civilizing Process*, i (New York, 1978), pp. xi–xvi.

20 M. Braddick, 'Introduction: The Politics of Gesture', in M. Braddick (ed.), *The Politics of Gesture: Historical Perspectives* (Oxford, 2009), p. 16.

21 Quoted in H. Hitchings, *Sorry! The English and Their Manners* (London, 2013), p. 166.

22 C. Hemphill, *Bowing to Necessities: A History of Manners in America, 1620–1860* (Oxford, 2002), p. 129.

23 Ibid., p. 158.

24 R. Collins, *Interaction Ritual Chains* (Princeton, 2005), p. 371.

25 E. Goffman, *Behavior in Public Places: Notes on the Social Organization of Gatherings* (New York, 1963), p. 4.

26 A. Kendon and A. Ferber, 'A Description of Some Human Greetings', in A. Kendon, *Conducting Interaction: Patterns of Behavior in Focused Encounters* (Cambridge, 2009), p. 199.

27 Goffman, *Behavior in Public Places*, p. 200.

28 Ö. Jónsson, 'Good Clean Fun: How the Outdoor Hot Tub Became the Most Frequented Gathering Place in Iceland' (2010), p. 247. Available at http://hdl.handle.net/1946/6754.

29 H. Garfinkel, 'Studies of the Routine Grounds of Everyday Activities', in J. Farganis (ed.), *Readings in Social Theory: The Classic Tradition to Post-Modernism* (New York, 2011), pp. 287–95.

30 E. Goffman, *Relations in Public: Microstudies of the Public Order* (New York, 1971), p. 83.

31 H. Sacks, *Lectures on Conversation*, i (Oxford, 1992), pp. 554–6.

32 Ibid., p. 549.

33 H. Sacks, *Lectures on Conversation*, ii (Oxford, 1992), pp. 197–9.

34 Collins, *Interaction Ritual Chains*, p. 243.

4. The Seductive Science of Body Language and the Quest for the Perfect Handshake

1 J. James, *The Body Language Bible* (London, 2008), p. 6.

2 A. Treviño (ed.), *Goffman's Legacy* (New York, 2003), p. 27.

3 E. Goffman, *The Presentation of Self in Everyday Life* (Edinburgh, 1956), p. 74.

4 D. Kahneman, *Thinking, Fast and Slow* (New York, 2011).

5 B. Englich, T. Mussweiler and F. Strack, 'Playing Dice with Criminal Sentences: The Influence of Irrelevant Anchors on Experts' Judicial Decision Making', *Personality and Social Psychology Bulletin* 32/2 (2006), pp. 188–200.

6 Quoted in Oliver Burkeman, 'Daniel Kahneman: "We're Beautiful Devices"', *Guardian* (14 November 2011). Available at www.theguardian.com/science/2011/nov/14/daniel-kahneman-psychologist.

7 S. Asch, 'Forming Impressions of Personality', *Journal of Abnormal and Social Psychology*, 41/3 (1946), pp. 258–90.

8 N. Ambady and R. Rosenthal, 'Half a Minute: Predicting Teaching Evaluations from Thin Slices of Nonverbal Behaviour and Physical Attractiveness', *Journal of Personality and Social Psychology*, 3/3 (1993), pp. 431–41.

9 J. Willis and A. Todorov, 'First Impressions: Making Up Your Mind after a 100-Ms Exposure to a Face', *Psychological Science*, 17/7 (2006).

10 See M. Hogg and G. Vaughan, *Social Psychology* (Harlow, 2014), pp. 82–109.

11 E. Jones and R. Nisbett, 'The Actor and the Observer: Divergent Perceptions and Causes of Behavior', in E. Jones et al., *Attribution: Perceiving the Causes of Behavior* (Morristown, NJ, 1972), pp. 79–84.

12 M. Leary and R. Kowalski, *Social Anxiety* (New York, 1995), pp. 6–10.

13 R. Kessler et al., 'Lifetime Prevalence and Age-of-Onset Distributions of DSM-IV Disorders in the National Comorbidity Survey Replication', *Archives of General Psychiatry* 62 (2005), pp. 593–602.

14 Leary and Kowalski, *Social Anxiety*, p. 53.

15 *Secrets of Body Language*, dir. James Millar [History Channel documentary, 2008].

16 C. Johnson, 'The 7%, 38%, 55% Myth', *Anchor Point*, 8/7 (1997), pp. 32–6.

17 A. Mehrabian and S. Ferris, 'Inference of Attitudes from Nonverbal Communication in Two Channels', *Journal of Consulting Psychology*, 31/3 (1967), pp. 248–58.

18 A. Mehrabian and M. Wiener, 'Decoding of Inconsistent Communications', *Journal of Personality and Social Psychology*, 6/1 (1967), pp. 109–14.

19 A. Mehrabian, *Silent Messages* (Belmont, CA, 1971), pp. 42–3.

20 M. Argyle et al., 'The Communication of Inferior and Superior Attitudes by Verbal and Nonverbal Signals', *British Journal of Social and Clinical Psychology*, 9 (1970), pp. 222–31.

21 J. Navarro, *What Every Body Is Saying: An Ex-FBI Agent's Guide to Speed-Reading People* (New York, 2008), p. 53.

22 Ibid., p. 88.

23 P. Ekman, *Emotions Revealed: Understanding Faces and Feelings* (London, 2003), p. 14.

24 Navarro, *What Every Body Is Saying*, pp. 206–8.

25 A. Summers, *The Arrogance of Power: The Secret World of Richard Nixon* (London, 2001), p. 206.

26 D. Greenberg, *Nixon's Shadow* (New York, 2003), pp. 70–1.

27 Quoted in Summers, *Arrogance of Power*, p. 208.

28 Navarro, *What Every Body Is Saying*, pp. 136–7.

29 'This Picture of Prince Will's Hand Was Taken after He Shook Hands with Modi, and It's Crazy', BuzzFeed [website]. Available at https://www.buzzfeed.com/sahilrizwan/56-psi-grip?utm_term=.qsqEAQ5RB#.helDQz2wq.

30 'Scientists Create Formula for Perfect Handshake', Newspress [website] (2010). Available at http://www.newspress.co.uk/public/ViewPressRelease.aspx?pr=23313.

31 W. Chaplin et al., 'Handshaking, Gender, Personality, and First Impressions', in *Journal of Personality and Social Psychology* 79/1 (2000), pp. 110–17.

32 G. Stewart et al., 'Exploring the Handshake in Employment Interviews', *Journal of Applied Psychology*, 93/5 (2008), pp. 1139–46.

33 Chaplin et al., 'Handshaking', p. 117.

5. When in Rome

1 K. Badt, *Greetings!* (Chicago, 1994), pp. 26–7.

2 J. Hevia, 'The Ultimate Gesture of Deference and Debasement: Kowtowing in China', in M. Braddick (ed.), *The Politics of Gesture: Historical Perspectives* (Oxford, 2009), p. 221.

3 *UNWTO Tourism Highlights, 2016 Edition* [UN World Tourism Organization publication], p. 4.

4 'Global Migration Trends Factsheet: 2015', International Organization for Migration [website]. Available at http://gmdac.iom.int/global-migration-trends-factsheet.

5 M. Ebsworth, J. Bodman and M. Carpenter, 'Cross-Cultural Realization of Greetings in American English', in S. Gass and J. Neu (eds), *Speech Acts across Cultures* (Berlin, 2006), p. 101.

6 Ibid., p. 100.

7 E. Tylor, *Primitive Culture: Researches into the Development of Mythology, Philosophy, Religion, Language, Art and Custom* (London, 1871), p. 1.

8 E. Tylor, *History of Mankind: Researches into the Early History of Mankind and the Development of Civilization* (London, 1865), p. 44.

9 D. Baynton, *Forbidden Signs: American Culture and the Campaign against Sign Language* (Chicago, 1996), pp. 36–55.

10 Quoted in K. Thomas, 'Introduction', in J. Bremmer and H. Roodenburg (eds), *A Cultural History of Gesture: From Antiquity to the Present Day* (Cambridge, 1991), p. 9.

11 Quoted in G. Stocking, *Race, Culture, and Evolution: Essays in the History of Anthropology* (Chicago, 1982), p. 148.

12 F. Boas, 'The Limitations of Comparative Anthropology', *Science*, 4 (1896), pp. 901–8.

13 F. Boas, *The Central Eskimo* (Washington DC, 1888), p. 609.

14 A. Kroeber, 'The Superorganic', *American Anthropologist*, 19/2 (1917), p. 163.

15 R. Benedict, *Patterns of Culture* (New York, 1989), p. 53.

16 R. Birdwhistell, *Kinesics and Context* (Philadelphia, 1970), p. 32.

17 A. Halberstadt, 'Toward an Ecology of Expressiveness: Family Socialization in Particular and a Model in General', in R. Feldman and B. Rimé (eds), *Fundamentals of Nonverbal Behavior* (New York, 1991), pp. 131–2.

18 K. Krys et al., 'Be Careful Where You Smile: Culture Shapes Judgments of Intelligence and Honesty of Smiling Individuals', *Journal of Nonverbal Behavior*, 40/2 (June 2016), pp. 101–16.

19 E. Hall, *The Hidden Dimension* (New York, 1969), p. 111.

20 *Secrets of Body Language*, dir. James Millar [History Channel documentary, 2008].

21 M. Knapp and J. Hall, *Nonverbal Communication in Human Interaction* (Boston, 2010), p. 464.

22 Ibid., pp. 354–5.

23 D. Foster, *The Global Etiquette Guide to Europe* (New York, 2000), p. 20.

24 K. Fox, *Watching the English: The Hidden Rules of English Behaviour* (London, 2004), p. 52.

25 'When Volcanic Ash Stopped a Jumbo at 37,000ft', BBC News [website] (15 April 2010). Available at http://news.bbc.co.uk/1/hi/magazine/8622099.stm.

26 'Why We Always Warm to the Weather', *Daily Telegraph* (1 October 2011). Available at http://www.telegraph.co.uk/news/weather/8801186/Why-we-always-warm-to-the-weather.html.

27 Fox, *Watching the English*, pp. 44–6.

28 Quoted in H. Hitchings, *Sorry! The English and Their Manners* (London, 2013), p. 50.

29 See K. Stark, *When Strangers Meet: How People You Don't Know Can Transform You* (New York, 2016), pp. 58–9; J. Moran, *Shrinking Violets: A Field Guide to Shyness* (London, 2016), p. 85.

30 Ibid., pp. 58–9.

31 I. Youssouf et al., 'Greetings in the Desert', *American Ethnologist*, 3/4 (November 1976), pp. 797–824. **32** E. Kezilahabi, 'A Phenomenological Interpretation of Kerebe Greetings', *Journal of African Cultural Studies*, 14/2 (December 2001), pp. 181–92.

33 J. Barlow and J. Nadeau, *The Bonjour Effect: The Secret Codes of French Conversation Revealed* (London, 2016), p. 269.

34 A. Duranti, 'Universal and Culture-Specific Properties of Greetings', *Journal of Linguistic Anthropology*, 7/1 (1997), pp. 63–97.

35 See Dean Foster's *Global Etiquette Guide* series.

36 Foster, *Global Etiquette Guide to Europe*, p. 4.

37 B. Parry, *Tribe: Adventures in a Changing World* (London, 2007), p. 316.

6. The Genital Grab and the Evolutionary Origins of Greetings

1 G. Davies, 'The Significance of the Handshake Motif in Classical Funerary Art', *American Journal of Archaeology*, 89/4 (1985), pp. 627–40.

2 C. Darwin, *The Descent of Man* (1871; London, 2004), p. 676.

3   See N. Tinbergen, *The Study of Instinct* (Oxford, 1976).

4   D. Cheney and R. Seyfarth, *Baboon Metaphysics: The Evolution of the Social Mind* (Chicago, 2007), pp. 5–6.

5   Tinbergen, *Study of Instinct*, p. 100.

6   Ibid., p. 119.

7   R. Dawkins, *The Selfish Gene* (Oxford, 1989), p. 146.

8   E. Wilson, *Sociobiology: The New Synthesis* (Harvard, 1975), p. 547.

9   D. Morris, *Peoplewatching* (London, 2002), p. 69.

10 K. Lorenz, *King Solomon's Ring* (1949; London, 2002), p. 157.

11 K. Lorenz, *On Aggression* (London, 1966), p. 179.

12 Quoted in F. de Waal, *The Age of Empathy: Nature's Lessons for a Kinder Society* (London, 2010), p. 207.

13 D. Peterson, *Jane Goodall: The Woman Who Redefined Man* (Boston, 2008), p. 212.

14 H. Nicholls, 'When I Met Jane Goodall, She Hugged Me Like a Chimp', *Guardian* (3 April 2014).

15 See J. Goodall, *The Chimpanzees of Gombe: Patterns of Behavior* (Cambridge, MA, 1986).

16 Available at https://www.youtube.com/watch?v=YzC7MfCtkzo.

17 J. Goodall, *In the Shadow of Man* (New York, 2000), p. 12.

18 Morris, *Peoplewatching*, p. 380.

19 F. de Waal, *The Bonobo and the Atheist* (New York, 2013), p. 14.

20 See 'Home Video: Breakfast with Baby Bear', The Kind Life [Alicia Silverstone's blog] (23 March 2012). Available at http://thekindlife.com/blog/2012/03/home-video-breakfast-with-baby-bear/.

21 T. Anthoney, 'The Ontogeny of Greeting, Grooming, and Sexual Motor Patterns in Captive Baboons', *Behaviour*, 31/3 (1968), pp. 359–62.

22 I. Eibl-Eibesfeldt, *Love and Hate: The Natural History of Behavior Patterns* (New York, 1971), p. 149.

23 Ibid., pp. 116–18.

24 F. de Waal, *Peacemaking among Primates* (Harvard, 1990), pp. 43–4.

25 de Waal, *The Bonobo and the Atheist*, p. 68.

26 F. de Waal, *Our Inner Ape: The Best and Worst of Human Nature* (London, 2006), p. 215.

27 D. Brown, *Human Universals* (New York, 1991), pp. 130–42.

28 C. Darwin, Beagle Diary, 18 December 1832. Available at http://darwinbeagle.blogspot.co.uk/2007/12/18th-december-1832.html.

29 See M. Nowak, *Supercooperators: Beyond the Survival of the Fittest* (London, 2012).

30 R. Dunbar, 'The Social Brain Hypothesis', *Evolutionary Anthropology* 6 (1998), pp. 178–90; 'Co-evolution of neocortex size, group size and language in humans', *Behavioral and Brain Sciences*, 16/4 (1993), pp. 681–735; *Human Evolution* (London, 2014), pp. 66–74.

31 *The Family of Chimps*, dir. Bert Haanstra (1984). Available at: https://www.youtube.com/playlist?list=PLfdrVIKBKkdZCfII6rAhwqpChOWI4Mwis.

32 A. Zahavi, 'Indirect Selection and Individual Selection in Sociobiology: My Personal Views on Theories of Social Behaviour', *Animal Behaviour*, 65/6 (2003), pp. 859–63;

'Altruism as a Handicap – the Limitations of Kin Selection and Reciprocity', *Avian Biology*, 26/1 (1995), pp. 1–3.

33 M. Bekoff, *The Emotional Lives of Animals* (Novato, CA, 2007), p. 3.

34 Darwin, *Descent of Man*, pp. 129 and 680.

35 See de Waal, *The Bonobo and the Atheist*.

36 Goodall, *Chimpanzees of Gombe*, pp. 488–534.

37 Lorenz, *On Aggression*, pp. 152–6.

38 A. Vincent, 'A Role for Daily Greetings in Maintaining Seahorse Pair Bonds', *Animal Behaviour*, 49 (1995), pp. 258–60.

39 R. Dunbar, *Grooming, Gossip and the Evolution of Language* (London, 1996).

40 R. Provine, *Curious Behavior: Yawning, Laughing, Hiccupping, and Beyond* (Cambridge, MA, 2012), p. 57.

41 Lorenz, *On Aggression*, p. 117.

42 Morris, *Peoplewatching*, p. 142.

43 F. de Waal, *Chimpanzee Politics: Power and Sex among Apes* (Baltimore, 2007), p. 82.

44 M. Meggitt, *The Desert People: A Study of the Walbiri Aborigines of Central Australia* (Sydney, 1986), p. 262.

45 B. Smuts and J. Watanabe, 'Social Relationships and Ritualized Greetings in Adult Male Baboons', *International Journal of Primatology*, 11/2 (1990), pp. 147–52; J. Whitham and D. Maestripieri, 'Primate Rituals: The Function of Greetings between Male Guinea Baboons', *Ethology*, 109 (2003), pp. 847–59.

46 L. Workman and W. Reader, *Evolutionary Psychology* (Cambridge, 2014), p. 2.

47 A. Zahavi and A. Zahavi, *The Handicap Principle: A Missing Piece of Darwin's Puzzle* (Oxford, 1996).

48 Ibid., p. 6.

49 Smuts and Watanabe, 'Social Relationships and Ritualized Greetings'.

50 S. Perry, *Manipulative Monkeys: The Capuchins of Lomas Barbudal* (Cambridge, MA, 2008), p. 255.

51 J. Smith et al., 'Greetings Promote Cooperation and Reinforce Social Bonds among Spotted Hyaenas', *Animal Behaviour*, 81 (2011), pp. 401–15.

52 H. Driessen, 'Gesture Masculinity: Body and Sociability in Rural Andalusia', in J. Bremmer and H. Roodenburg (eds), *A Cultural History of Gesture: From Antiquity to the Present Day* (Cambridge, 1991), p. 244.

53 See 'Does a British Person Actually Like You?', British.com [blog] (24 February 2017). Available at https://britrish.com/2017/02/24/does-a-british-person-actually-like-you/.

54 F. de Waal, *Peacemaking among Primates* (Harvard, 1990), p. 79.

7. It's All in the Mind

1 See the transcript of Juan Mann's interview on the ABC series *Enough Rope with Andrew Denton*, broadcast 1 October 2007. Available at http://www.abc.net.au/tv/enoughrope/transcripts/s2045334.htm; 'Free-Hug Man Speaks Out', *Sydney Morning Herald* (28 September 2006); 'On the Trail of the Free Hugs Founder', *Guardian* (2 September 2012).

2 'A Hug from Amma', BBC News [website] (6 December 2007). Available at http://news.bbc.co.uk/1/hi/magazine/7130151.stm.

3 See, for example, C. Frith, 'The Social Brain?', *Philosophical Transactions*, 368 (2007), pp. 671–8; C. Frith and U. Frith, 'The Biological Basis of Social Interaction', *Current Directions in Psychological Science*, 10/5 (October 2001), pp. 151–5.

4 K. Kelleway, 'Uta Frith: "The Brain Is Not a Pudding; It Is an Engine"', *Guardian* (24 February 2013).

5 On mirror neurons see, for example, M. Iacoboni, 'Imitation, Empathy, and Mirror Neurons', *Annual Review of Psychology*, 60 (2009), pp, 653–70.

6 See C. Frith, 'How the Brain Creates Culture' [paper presented to the German Academy of Sciences Leopoldina, September 2013].

7 See C. Frith, 'The Role of Metacognition in Human Social Interactions', *Philosophical Transactions*, 367 (2012), pp. 2213–23.

8 A. Todorov et al., 'Social Judgments from Faces', *Current Opinion in Neurobiology*, 23 (2013), pp. 373–80; A. Todorov et al., 'Evaluating Face Trustworthiness: A Model-Based Approach', *Social Cognitive and Affective Neuroscience*, 3/2 (2008), pp. 119–27.

9 D. Kelly et al., 'Three-Month-Olds, but Not Newborns, Prefer Own-Race Faces', *Developmental Science*, 8/6 (November 2005), pp. F31–6.

10 A. Hart et al., 'Differential Response in the Human Amygdala to Racial Outgroup vs Ingroup Face Stimuli', *NeuroReport*, 11/11 (2000), pp. 2351–5.

11 S. Fiske, 'Intent and Ordinary Bias: Unintended Thought and Social Motivation Create Casual Prejudice', *Social Justice Research*, 17/2 (June 2004).

12 C. Darwin, *The Expression of the Emotions in Man and Animals* (1872; London, 2009), p. 286.

13 R. Crozier, 'The Puzzle of Blushing', *Psychologist*, 23 (May 2010), pp. 390–3.

14 N. Eisenberger, 'Does Rejection Hurt? An fMRI Study of Social Exclusion', *Science*, 302 (October 2003), pp. 290–2.

15 S. Asch, 'Studies of Independence and Conformity:I. A Minority of One against a Unanimous Majority', *Psychological Monographs: General and Applied*, 70/9 (1956).

16 R. Provine, *Curious Behavior: Yawning, Laughing, Hiccupping, and Beyond* (Cambridge, MA, 2012), p. 101.

17 E. Hess, 'The Role of Pupil Size in Communication', *Scientific American*, 233/5 (November 1975), pp. 110–19.

18 M. Bradley et al., 'The Pupil as a Measure of Emotional Arousal and Autonomic Activation', *Psychophysiology* 45/4 (2008), pp. 602–7.

19 P. Ekman, *Emotions Revealed: Understanding Faces and Feelings* (London, 2003), pp. 9–14.

20 P. Ekman and W. Friesen, *Unmasking the Face: A Guide to Recognizing Emotions from Facial Expression* (Cambridge, MA, 2003), pp. 23–4.

21 D. Matsumoto, 'Cultural Influences on Facial Expressions of Emotion', *Southern Communication Journal*, 56 (1989), pp. 128–37.

22 Quoted in Ekman, *Emotions Revealed*, p. 206.

23 M. Frank and P. Ekman, 'Physiological Effects of the Smile', *Directions in Psychiatry*, 16/25 (December 1996), pp. 1–8.

24 F. Strack, 'Inhibiting and Facilitating Conditions of the Human Smile: A

Nonobtrusive Test of Facial Feedback Hypothesis', *Journal of Personality and Social Psychology*, 54/5 (May 1988), pp. 768–77.

25 D. Kennedy et al., 'Personal Space Regulation by the Human Amygdala', *Nature Neuroscience*, 12/10 (2009), pp. 1226–7.

26 J. Suvilehto et al., 'Topography of Social Touching Depends on Emotional Bonds between Humans', *Proceedings of the National Academy of Sciences*, 112/45 (November 2015), pp. 13811–16.

27 I. Morrison et al., 'The Skin as a Social Organ', *Experimental Brain Research*, 204/3 (September 2009), pp. 305–14.

28 Provine, *Curious Behavior*, p. 173.

29 D. Linden, *Touch: The Science of Hand, Heart, and Mind* (London, 2015), pp. 22–7.

30 M. Knapp and J. Hall, *Nonverbal Communication in Human Interaction* (Boston, 2010), p. 262.

31 M. Lynn et al., 'Reach Out and Touch Your Customer', *Cornell Hotel and Restaurant Administration Quarterly*, 39 (1998), pp. 60–5.

32 J. Levav and J. Argo, 'Physical Contact and Financial Risk Taking', *Psychological Science*, 21/6 (April 2010), pp. 1–7.

33 Suvilehto et al., 'Topography of Social Touching'.

34 M. Kraus et al., 'Tactile Communication, Cooperation, and Performance: An Ethological Study of the NBA', *Emotion*, 10/5 (2010), pp. 745–9.

35 A. Theodoridou et al., 'Oxytocin and Social Perception: Oxytocin Increases Perceived Facial Trustworthiness and Attractiveness', *Hormones and Behaviour*, 56/1 (June 2009), pp. 128–32; M. Kosfeld, 'Oxytocin Increases Trust in Humans', *Nature*, 453 (2005), pp. 673–6.

36 See P. Zak, 'The Neurobiology of Trust', *Scientific American*, 298/6 (2008), pp. 88–95; *The Moral Molecule: The New Science of What Makes Us Good or Evil* (London, 2013).

37 The words of one customer review on the Amazon.com page for OxyLuv. Available at https://www.amazon.com/OxyLuv-Oxytocin-unnatural-preservatives-fillers/dp/B00MEV8BMM.

38 P. Zak, 'The Power of a Handshake: How Touch Sustains Personal and Business Relationships', *Huffington Post* (25 May 2011). Available at http://www.huffingtonpost.com/paul-j-zak/the-power-of-a-handshake_b_129441.html.

39 Ibid.

40 See, for example, G. Nave et al., 'Does Oxytocin Increase Trust in Humans? A Critical Review of Research', *Perspectives on Psychological Science*, 10/6 (2015), pp. 772–89.

41 J. Dabbs and F. Bernieri, 'Going on Stage: Testosterone in Greetings and Meetings', *Journal of Research in Personality*, 35 (2001), pp. 27–40.

42 A. Cuddy et al., 'Power Posing: Brief Nonverbal Displays Affect Neuroendocrine Levels and Risk Tolerance', *Psychological Science*, 21/10 (2010), pp. 1363–8.

43 E. Ranehill et al., 'Assessing the Robustness of Power Posing: No Effect on Hormones and Risk Tolerance in a Large Sample of Men and Women', *Psychological Science*, 26/5 (2015), pp. 653–6.

44 A. Cuddy, *Presence: Bringing Your Boldest Self to Your Biggest Challenges* (New York, 2015), pp. 72–5.

45 L. Williams and J. Bargh, 'Experiencing Physical Warmth Promotes Interpersonal Warmth', *Science*, 322 (2008), pp. 606–7.

46 D. Chen and J. Haviland-Jones, 'Human Olfactory Communication of Emotion', *Perceptual and Motor Skills*, 91 (2000), pp. 771–81.

47 J. Groot et al., 'Chemosignals Communicate Human Emotions', *Psychological Science*, 23/11 (2012), pp. 1417–24.

48 I. Frumin et al., 'A Social Chemosignaling Function for Human Handshaking', *eLife* (March 2015). Available at https://elifesciences.org/content/4/e05154/article-info.

49 S. Blakemore, 'The Social Brain in Adolescence', *Nature Reviews Neuroscience*, 9 (April 2008), pp. 267–77; L. Somerville et al., 'The Medial Prefrontal Cortex and the Emergence of Self-Conscious Emotion in Adolescence', *Psychological Science*, 24/8 (2013), pp. 1554–62.

50 L. Spear, 'The Adolescent Brain and Age-Related Behavioral Manifestations', *Neuroscience and Behavioral Reviews*, 24 (2000), pp. 417–63.

51 For a summary of this line of argument, see Simon Baron-Cohen, *The Essential Difference: Men, Women and the Extreme Male Brain* (London, 2003).

8. Forget about Chimpanzees

1 A. Whiten et al., 'Cultures in Chimpanzees', *Nature*, 399 (June 1999), pp. 682–5.

2 R. Wrangham et al., 'Distribution of a Chimpanzee Social Custom Is Explained by Matrilineal Relationship Rather than Conformity', *Current Biology*, 26/22 (November 2016), pp. 1–5.

3 R. Foley and M. Mirazón Lahr, 'The Evolution of Diversity in Cultures', *Philosophical Transactions: Biological Sciences*, 366/1567 (2011), pp. 1080–9.

4 M. Cook, *A Brief History of the Human Race* (London, 2005), p. 68.

5 K. Thomas, 'Introduction', in J. Bremmer and H. Roodenburg (eds), *A Cultural History of Gesture: From Antiquity to the Present Day* (Cambridge, 1991), p. 11.

6 E. Evans-Pritchard, *Anthropology and History: A Lecture* (Manchester, 1961), pp. 13–14.

7 W. Frijhoff, 'The Kiss Sacred and Profane: Reflections on a Cross-Cultural Confrontation', in J. Bremmer and H. Roodenburg (eds), *A Cultural History of Gesture: From Antiquity to the Present Day* (Cambridge, 1991), p. 224.

8 See, for example, J. McNeill and W. McNeill, *The Human Web: A Bird's Eye View of Human History* (New York, 2003); Cook, *A Brief History of the Human Race*; Y. Harari, *Sapiens: A Brief History of Humankind* (London, 2015).

9 C. Koslofsky, 'The Kiss of Peace in the German Reformation', in K. Harvey (ed.) *The Kiss in History* (Manchester, 2005), p. 19.

10 M. Bormann, *Hitler's Table Talk, 1941–1944* (London, 2000), pp. 172–3.

11 D. Arnold, 'Salutation and Subversion: Gestural Politics in Nineteenth-Century India', in M. Braddick (ed.), *The Politics of Gesture: Historical Perspectives* (Oxford, 2009), p. 193.

12 M. Braddick, 'Introduction: The Politics of Gesture', in M. Braddick (ed.), *The Politics of Gesture: Historical Perspectives* (Oxford, 2009), pp. 28–9.

13 K. Sennefelt, 'The Politics of Hanging Around and Tagging Along: Everyday

Practices of Politics in Eighteenth Century Stockholm', in M. Braddick (ed.), *The Politics of Gesture: Historical Perspectives* (Oxford, 2009), p. 178.

14 Braddick, 'Introduction: The Politics of Gesture', p. 28.

15 Ibid., pp. 28–9.

16 Ibid., p. 16.

17 J. Walter, 'Gesturing at Authority: Deciphering the Gestural Code of Early Modern England', in M. Braddick (ed.), *The Politics of Gesture: Historical Perspectives* (Oxford, 2009), p. 114.

18 H. Roodenburg, 'The "Hand of Friendship": Shaking Hands and Other Gestures in the Dutch Republic', in J. Bremmer and H. Roodenburg (eds), *A Cultural History of Gesture: From Antiquity to the Present Day* (Cambridge, 1991), pp. 171–4.

19 K. Thomas, 'Afterword', in K. Harvey (ed.) *The Kiss in History* (Manchester, 2005), p. 194.

20 See Thomas, 'Afterword', pp. 198–9.

21 S. Kirshenbaum, *The Science of Kissing: What Our Lips Are Telling Us* (New York, 2011), p. 52.

22 S. Mela and D. Whitworth, 'The Fist Bump: A More Hygienic Alternative to the Handshake', *American Journal of Infection Control*, 42 (2014), pp. 916–17.

9. Back to the Future or a Brave New World?

1 R. Collins, *Interaction Ritual Chains* (Princeton, 2005), p. 63.

2 S. Greenfield, *Mind Change: How Digital Technologies Are Leaving Their Mark on Our Brains* (London, 2014), pp. 141–2.

3 F. Fukuyama, *The End of History and the Last Man* (New York, 1992).

4 D. Harmon and J. Loh, 'The Index of Linguistic Diversity: A New Quantitative Measure of Trends in the Status of the World's Languages', *Language Documentation and Conservation*, 4 (2010), p. 97.

5 S. Huntington, *The Clash of Civilizations and the Remaking of World Order* (New York, 1997).

6 Fukuyama, *End of History*, pp. 330–1.

Goodbye!

1 H. Berry, 'Lawful Kisses? Sexual Ambiguity and Platonic Friendship in England c.1660–1720', in K. Harvey (ed.) *The Kiss in History* (Manchester, 2005), pp. 62–75.

2 K. Harvey, 'Introduction' and 'Afterword', in K. Harvey (ed.) *The Kiss in History* (Manchester, 2005), pp. 10, 196.

3 Thomas, 'Afterword', p. 195.

4 R. Firth, 'Verbal and Bodily Rituals of Greeting and Parting', in J. La Fontaine (ed.), *The Interpretation of Ritual: Essays in Honour of A. I. Richards* (London, 1972), pp. 26–7.

5 F. de Waal, *Are We Smart Enough to Know How Smart Animals Are?* (London, 2016), p. 3.

6 Stark, *When Strangers Meet*, p. 73.

Index